Motor Vehicle Engineering

Level 3

D0421004

Tom Denton, BA, AMSAE, AMIRTE, Cert. Ed.

THOMSON

LEARNING

Australia • Canada • Mexico • Singapore • Spain • United Kingdom • United States

THOMSON

LEARNING

Motor Vehicle Engineering – Level 3

Copyright   Tom Denton 1998

The Thomson Learning logo is a registered trademark used herein under licence.

For more information, contact Thomson Learning, Berkshire House, 168–173 High Holborn, London, WC1V 7AA or visit us on the World Wide Web at:
http://www.thomsonlearning.co.uk

British Library Cataloguing-in-Publication Data
A catalogue record for this book is available from the British Library

ISBN 1-86152-744-6

First edition published 1998 by Macmillan Press Ltd

Reprinted 2001 by Thomson Learning

Printed and bound by Antony Rowe Ltd, Eastbourne

Contents

		page
Preface		iv
Acknowledgements		v
UPK statements		vi
1	**Introduction and welcome back!**	1
2	**Health and safety**	12
3	**Tools and equipment**	20
4	**Engines**	35
5	**Engine management systems**	64
6	**Transmission**	121
7	**Suspension**	142
8	**Steering**	156
9	**Brakes**	176
10	**Electrical systems and circuits**	184
11	**Comfort and safety**	220
12	**Diagnostics**	243
13	**Augmentation**	256
14	**Valeting**	269
15	**Coaching individuals and key skills**	280
16	**Conclusions and what next?**	287
	Index	289

Preface

I still have the following theory:

'A car has an engine and some other bits to make the wheels go round. When it breaks down or needs a service it will, in most cases, be taken to a garage for repairs. The garage will operate normal company systems to fix the car and make a profit.'

What I mean by this is that different qualifications and courses come and go, but my theory will still be good! Therefore, what you will find in this book (and the level two version) will be relevant for quite some time, no matter what.

I still enjoy learning new things about vehicles and the motor trade in general. I hope you enjoy your career as much as I have, and still do today.

I would very much welcome further comments from lecturers and students. You can visit my website at:

> http://ourworld.compuserve.com/homepages/tom_denton/

I will be offering technical updates, advice about using this book for new or changed qualifications, and information about future titles. I will also be supplying new and updated learning tasks. Free software, shareware programs and other files are available for downloading. All suggestions are welcome!

Tom Denton 1998

Acknowledgements

Thanks are due to the following companies who have supplied information and/or permission to reproduce photographs and diagrams:

AA Photo Library   5.4, 5.13, 6.3, 6.5, 6.6, 6.12, 6.17, 6.18, 6.19, 6.20, 7.11, 9.4, 9.5, 9.6

Autodata Ltd.   1.1

Autoglym Ltd.   14.3, 14.5, 14.6, 14.8

BMW Cars UK Ltd.   4.1, 4.33, 4.34, 4.35, 4.36, 4.37

Fiat Cars UK Ltd.   11.2

Fluke Instruments   3.7

Ford Motor Company Ltd.   2.4, 6.1, 6.2, 6.7, 6.8, 6.10, 6.14, 6.16, 6.22, 7.1, 7.2, 7.6, 7.7, 7.9, 7.10, 7.12, 7.13, 7.14, 7.15, 7.16, 7.17, 7.18, 7.20, 7.21, 7.22, 7.23, 7.24, 7.26, 7.27, 7.28, 7.29, 7.30, 8.2, 8.9, 8.10, 8.11, 8.12, 8.17, 8.18, 8.19, 8.21, 8.22, 8.23, 8.30, 8.31, 8.32, 8.33, 8.35, 11.4, 12.2, 14.1

GenRad Ltd.   3.16, 3.18

Girling Ltd.   9.7, 9.8

Goodyear Tyre and Rubber Co.   8.36

HMSO   13.6, 13.7, 13.8, 13.9

Honda Cars UK Ltd.   5.47

Johnson Matthey UK Ltd.   5.52

Karcher Ltd.   14.4, 14.7

Lucas Ltd.   3.14, 5.5, 5.6, 5.14, 5.21, 5.28, 5.29, 5.30, 10.10, 10.14, 10.19, 10.21, 10.31

Peugeot UK Ltd.   2.5, 13.1, 13.10, 13.11

Pioneer Radio Ltd.   13.3, 13.4, 13.5

Renault UK Ltd.   2.6, 11.1

Robert Bosch GmbH   3.17, 5.13, 5.35, 5.36, 5.37, 5.38, 5.39, 5.40, 5.41, 5.42, 5.43, 5.44, 5.51, 6.14, 9.9, 10.5, 10.20

Robert Bosch Press Photos   3.15, 4.31, 5.1, 5.9, 5.11, 5.16, 5.17, 5.18, 5.20, 5.22, 5.24, 5.25, 5.26, 5.27, 5.31, 5.32, 5.33, 5.34, 5.40, 5.45, 5.48, 5.53, 5.54, 10.3, 10.8, 11.5

Rover Cars Ltd.   4.3, 4.7, 4.11, 4.17, 4.22, 4.23, 4.30, 5.7, 5.46, 6.9, 6.11, 6.13, 6.21, 6.23, 6.24, 6.25, 6.26, 8.20, 8.25, 9.1, 9.2, 9.3, 10.2, 10.28, 10.33, 10.43, 11.3, 11.11, 11.16, 11.17, 11.18, 11.19, 11.20, 11.22

Saab Cars UK Ltd.   4.8, 4.9, 4.10, 5.10, 8.26, 10.41

Snap-on Tools Inc.   3.8

Toyota Cars UK Ltd.   7.9, 13.12, 13.13, 13.15

Unipart Group Ltd.   10.26, 10.29

VDO Instruments   10.44

Witter Towing Brackets Ltd.   13.16

Thanks again to the listed companies. If I have used any information or mentioned a company name not listed here, my apologies and acknowledgements. Last, but by no means least, thank you yet again to my family: Vanda, Malcolm and Beth.

UPK statements

Please note:
This section refers you to the most relevant pages or general area relating to the UPK statements. Many of these statements overlap, so you will have to refer to several areas. The shaded sections are page references in the level two book.

Unit no.	Unit title and UPK statements	Page numbers
A1–G	**Contract with customers to provide for their vehicle needs**	
A1.1	Vehicle test programmes	227, 230
	Courses of action	245
	Customer relations	244, 245, 252
	Estimating costs and times	247
	Company services	241, 243, 244
	Resources available	243
A1.2	Requirements of a contract	255
	Warranty schemes	249
	Recording systems	231, 235
	Documentation requirements	248, 249
	Limits of authority	253
A6–G	**Reinstate the cleanliness of the vehicle**	
A6.1	Properties of exterior cleaning materials	270, 273
	Use of tools and equipment	276, 277
	Safety precautions	270, 271
A6.2	Properties of vehicle interior cleaning materials	277
	Use of tools and equipment	277, 278
	Safety precautions	270, 271
A7–L	**Routinely service the vehicle to maintain optimum performance**	
A7.1	Reasons for regular or special vehicle inspections	227, 233
	General legislative requirements for road vehicles	10, 17, 164, 234
	Interpretation of vehicle data	229, 235
	Purposes of vehicle inspection records	231
	Requirements of customer contracts	255
A7.2	Reasons for regular adjustment	233
	Interpretation of data	229, 235
	Effects of incorrect adjustments	239
	Identification codes and grades of lubricants	116

Unit no.	Unit title and UPK statements	Page numbers
	Use of parts lists and identification codes	246
	Reasons for servicing records	230, 235
A8–L2	**Identify faulty components/units which affect system performance**	
A8.1	Planned systematic procedures for testing	62, 77, 89, 98, 112, 123, 137, 151, 164, 180, 212, 225, 230
	Interpretation of test data and instructions	62, 77, 89, 98, 112, 123, 137, 151, 164, 180, 212, 225, 229, 230
	Use of test equipment	33, 62, 77, 89, 98, 112, 123, 137, 151, 164, 180, 212, 225, 230
A8.2	Operating principles of vehicle systems	36, 45, 64, 80, 92, 100, 115, 126, 139, 153, 166, 182, 215
	Interpretation of test results and vehicle data	62, 77, 89, 98, 112, 123, 137, 151, 164, 180, 212, 225, 229, 230
	Identification of defective units/components	62, 77, 89, 98, 112, 123, 137, 151, 164, 180, 212, 225, 230
	Possible causes of faults	62, 77, 89, 98, 112, 123, 137, 151, 164, 180, 212, 225, 230
	Replacement procedures	36, 45, 64, 80, 92, 100, 115, 126, 139, 153, 166, 182, 215
A8–L3	**Identify faults which affect system performance**	
A8.1	Planned systematic procedures for testing	243, 244
	Interpretation of test data and instructions	245, 252
	Accuracy of test equipment	20, 23, 31
	Use of test equipment	20, 30, 32
A8.2	Operating principles of vehicle systems	35, 64, 121, 142, 156, 176, 184, 220
	Interpretation of test results and vehicle data	243, 244
	Identification of defective units/components	243, 244
	Possible causes of faults	245
	Rectification and repair procedures	244
A9–L2	**Remove and replace vehicle components/units**	
A9.1	Constructional features of vehicle systems	36, 45, 64, 80, 92, 100, 115, 126, 139, 153, 166, 182, 215
	Safe practices and procedures	13, 15, 16, 18
	Use of tools and equipment	32, 34
	Interpretation of technical information	229, 235
A9.2	Constructional features of vehicle systems	36, 45, 64, 80, 92, 100, 115, 126, 139, 153, 166, 182, 215

Unit no.	Unit title and UPK statements	Page numbers
	Operating characteristics of vehicle systems	227
	Use of tools and equipment	32, 34
	Interpretation of inspection results and vehicle data	62, 77, 89, 98, 112, 123, 137, 151, 164, 180, 212, 225, 230
A9.3	Constructional features of vehicle systems	36, 45, 64, 80, 92, 100, 115, 126, 139, 153, 166, 182, 215
	Operating characteristics of vehicle systems	36, 45, 64, 80, 92, 100, 115, 126, 139, 153, 166, 182, 215
	Interpretation of technical information	229, 235
	Methods of reassembly	36, 45, 64, 80, 92, 100, 115, 126, 139, 153, 166, 182, 215
	Use of tools and equipment	32, 34
	Properties of jointing materials	30, 31
A9.4	Test procedures	62, 77, 89, 98, 112, 123, 137, 151, 164, 180, 212, 225, 230
	Road safety	17, 232, 238
	Evaluation of test information	62, 77, 89, 98, 112, 123, 137, 151, 164, 180, 212, 225, 230
	Statutory requirements for vehicles	10, 16, 17, 164, 234
	Customer contract requirements	255
A9–L3	**Rectify faults in vehicle systems**	
A9.1	Constructional features of vehicle systems	35, 64, 121, 142, 156, 176, 184, 220
	Safe practices and procedures	12
	Use of tools and equipment	20
	Interpretation of technical information	24, 32, L2–229, 235
A9.2	Constructional features of vehicle systems	35, 64, 121, 142, 156, 176, 184, 220
	Operational characteristics of vehicle systems	35, 64, 121, 142, 156, 176, 184, 220
	Accuracy of measuring and inspection equipment	23, 31
	Use of tools and equipment	20, 23, 31
	Interpretation of inspection results and vehicle data	243, 244
A9.3	Methods of repair and reclamation	35, 64, 121, 142, 156, 176, 184, 220
	Limits, fits and tolerances	23, 31
	Operational characteristics of vehicle systems	35, 64, 121, 142, 156, 176, 184, 220
	Constructional features of vehicle systems	35, 64, 121, 142, 156, 176, 184, 220
	Rectification of the cause(s) of faults	243, 244

Unit no.	Unit title and UPK statements	Page numbers
A9.4	Constructional features of vehicle systems	35, 64, 121, 142, 156, 176, 184, 220
	Operational characteristics of vehicle systems	35, 64, 121, 142, 156, 176, 184, 220
	Interpretation of technical information	243, 244
	Methods of reassembly	L2–36, 45, 64, 80, 92, 100, 115, 126, 139, 153, 166, 182, 215
	Use of tools and equipment	20, 23, 31
	Properties of jointing materials	L2-30, 31
A9.5	Test procedures	243, 244, 245, 252
	Road safety	L2–17, 232, 238
	Evaluation of test information	L2–62, 77, 89, 98, 112, 123, 137, 151, 164, 180, 212, 225, 230
	Statutory requirements for vehicles	L2–10, 16, 17, 164, 234
	Customer contract requirements	L2–255
A10–LH	**Augment vehicles to meet customer requirements**	
A10.1	Types of mechanical/electrical augmentation	256, 257, 258
	Benefits of mechanical/electrical augmentation	265, 266
	Disadvantages of mechanical/electrical augmentation	265, 266
	Possible legislative implications of augmentations	268
A10.2	Interpretation of technical instructions	257, 267
	Interpretation of customers' instructions	257
	Use of tools and equipment	20, 23, 31
A11–L	**Prepare new, used or repaired vehicles for customer use**	
A11.1	Operation of vehicle systems	227
	Test and inspection procedures	230, 234
	Requirements of recording systems	231, 235
	Road safety	17, 232, 238
A11.2	Interpretation of vehicle specifications	229
	Requirements of recording systems	235
A12–G	**Maintain effective working relationships**	
A12.1	Customer requirements and expectations	248, 252, 254
	Importance of the customer	245, 252, 256
	Customer relations techniques	245, 254, 256
	Dealing with complaints	255, 256
	Security of customers' property	257
A12.2	Lines of communication	244, 245, 253
	Levels of authority	253
	Interdependence of all staff	242, 252
	Company image	242

Unit no.	Unit title and UPK statements	Page numbers
A13–G	Maintain the health, safety and security of the working environment	
A13.1	Health and safety regulations	10, 11, 12, 13, 22
	Environmental protection regulations	11, 12, 23
	Duties of the individual	13, 15, 16
A13.2	Safety procedures	15, 80
	Fire regulations	19, 20
	Reporting requirements	20
A13.3	Cleaning requirements	21
	Properties of cleaning materials	21
	Safety regulations	16, 22

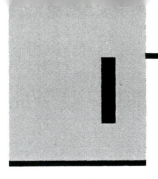

Introduction and welcome back!

1 Introduction

Start here! Hello! This book will help you understand and develop the background knowledge needed for your NVQ **level 3**, or indeed, any other qualification in motor vehicle engineering. The term used to describe this is 'underpinning knowledge' or 'UPK' for short. Motor vehicle engineering is a very practical subject. This book does not try to cover the practice in detail. However, in order to carry out practical work to national standards, a good background knowledge is essential.

To gain your qualification, you will also need to study the level 2 book. If you have already done so then you will know how this book works. This chapter is quite similar to the level 2 version, but with some additions. However, I still suggest you scan over it again to make sure you know the basics of what you must do. The main chapters that follow will assume that you have completed your study of the level 2 book.

You can study the contents of this book in any order you like, but it is not intended to be a 'self learning' study pack. It is best used in conjunction with a course of study. You will also find it a useful reference for looking up information as required. Figure 1.1 shows a popular information source: data books.

Figure 1.1 Data books are an excellent source of information

To obtain an NVQ, you will need to show your assessor that you can do the work and that you have the necessary UPK. For other qualifications, you must pass written and practical examinations. This book will be useful either

Figure 1.2 Vehicle position symbols – to make sure you get the right information

way. Because our chosen subject is now so wide ranging, you will also need to look at other sources of information. These other sources will be some or all of the following:

- your college teacher or assessor
- data books
- workshop manuals
- other textbooks, magazines and newspapers
- CD ROM information
- videos
- your workplace supervisor or experienced work mates

The fine details of the UPK you need will vary depending on your particular situation at work. However, the basic facts are the same for all motor vehicle systems and workshop operation methods.

Good luck with your work and studies, and remember: in order to get anything out of it, you will have to put something in! As a final incentive, just check the newspaper job adverts and note that the top technicians earn very good money. This may be several years away yet, but it is worth putting in the effort now.

Terminology

NCVQ	National Council for Vocational Qualifications
MITSC	Motor Industry Training Standards Council. The people who set our standards, sometimes called the lead body
Modern Apprenticeship	This scheme requires all the NVQ units to be achieved as well as a number of key skills. It is designed to formalise apprentice training to NVQ standards. A contract is required between the local Training and Enterprise Council (TEC), the employer and the employee
NVQ	National Vocational Qualifications
UPK	Underpinning knowledge – most of which you will learn from this book and the level 2 version.
Evidence	Material to prove you can do something
Assessor	The person who will assess and advise you in gaining your NVQ
Portfolio	The collection of evidence which is cross-referenced. Usually a ring binder
Standards	The lists of units, elements and performance criteria, set by MITSC
Units	Main parts required building up to an NVQ. About six or seven units is normal
Elements	Each unit is split into elements to further explain what is needed
Performance criteria	Each element has many performance criteria, which state exactly what you must be able to do
Range statements	These describe the range of subjects or systems you will cover in the performance criteria
Other qualifications	Traditional qualifications usually involve passing written examinations and practical tests

What is an NVQ Level 3?

National Vocational Qualifications (NVQs) were designed to meet the modern and ever changing needs of our industry. They are also important in that a national standard has been set so that we are all working towards the same thing!

The aim of NVQs is to enable you to do your job. This is, of course, what an employer wants and therefore will help you get a good job.

Figure 1.3 Motor vehicle engineering is a good trade to work in

'Vocational' means related to the work you do, but the qualification still requires a large amount of understanding about motor vehicle technology. Working in one particular garage or dealership specialising in one make of car, you will probably become quite good at fault finding the fuel or ignition system. This is because you know the systems well, have seen it all before and have all the information you need. When a different type of car comes in for you to repair a similar fault, you need to understand the underpinning principles to be able to carry out your work and earn the good money you deserve.

In the motor industry, NVQs and other qualifications exist at different levels. These reflect the level of work you are or will be doing. I hope by now you have completed, or almost completed, your level 2. If so you are well on your way – keep it up.

- NVQ Level 1: basic tasks mostly under supervision.
- NVQ Level 2: diagnose and repair faults to unit level.
- NVQ Level 3: diagnose and repair detailed faults plus some degree of supervision of others.
- NVQ Level 4/5: management and or advanced diagnostic skills.

An NVQ is made up of a number of units. Each unit has a number of elements (usually 2, 3 or 4). These elements have lists of 'performance criteria'. These are at the heart of your NVQ: they specify what you must be able to do. There is also a list of the UPK and the range of vehicles and/or systems as appropriate to each unit.

The qualification is awarded when you have collected enough evidence to show that you can meet all the requirements of all the units at your level. It is possible to gain credit for individual units, but most students aim at the complete qualification. Many students find the language used in the standards confusing – well it is! The reason – is to keep the standards up to date with modern practices and different work situations without having to constantly change the wording. So we are stuck with it! If you have completed your level 2 then you will have got the hang of it by now. It is just like learning about motor vehicle engineering, or any other new subject. Once you have learnt the new words, then the rest of it becomes much easier. This is why the 'terminology' sections are included at the start of each main subject area.

How does a level 3 differ from a level 2?

The key difference is that level 3 is described as '**repair**' whereas level 2 is described as '**replace**'. This means you are expected to have a more detailed understanding of the technology and repair procedures. The most important aspect is still being able to 'do the job'.

You will also be expected to take a greater role in the supervision of others and making decisions on your own. This is a very big responsibility, but you have already proved you can do a lot by getting this far. As I said before, keep it up!

LEARNING TASKS

➡ Look back at the key words. Explain each one to a friend, and/or write out a short description to keep as evidence.

➡ Consider what it means to make decisions about car repairs, in a situation where any further problems will be referred back to you.

➡ Look at a workshop manual and make a note of two tasks, one you would describe as 'repair', and the other as 'replace'.

---

2 Motor vehicle NVQ units and what they mean

Introduction

In this section I give you my interpretation of the NVQ units but you should check with your assessor who will be able to be more specific about details as they relate to you. Each unit covers many more situations than the examples given here, but these will serve as a useful guide. Depending on the qualification you are aiming at, you may not need to achieve all the units.

NVQ units – a very simplified interpretation

Unit	Interpretation	Example
Unit A1–G		
Contract with customers to provide for their vehicle needs	You must be able to meet customers, deal with their problems, answer questions and understand the importance of a customer contract with your company	Meet a customer at reception, find out what the problem is, quote a price and book the car in to the workshop
Unit A8–L2		
Identify faulty components/units which affect system performance	When a car has a problem, you must be able to find the cause or reason	The car you are working on makes a grinding or rumbling noise, but only when going round sharp left hand bends. It's probably a wheel bearing, but which one?
Unit A9–L2		
Remove and replace vehicle components/units	This is the main part of our work, together with the previous unit. Remove faulty parts and replace with new. It means following repair procedures such as in a workshop manual	Renew the wheel bearing identified above
Unit A11–L		
Prepare new, used or repaired vehicles for customer use	Inspect a vehicle and make sure it is safe, clean and suitable for a customer to use	Carry out a pre-delivery inspection (PDI) or a used vehicle inspection

Unit	Interpretation	Example
Unit A12–G Maintain effective working relationships	Get on well at work with your customers, workmates, supervisor and boss	You must be able to accept instructions and work well as part of a team
Unit A13–G Maintain the health, safety and security of the working environment	Deal with toxic materials in the correct way, follow sensible safety procedures and look after customers' property	Clean brakes with proper solvents, use axle stands after jacking up a vehicle and lock the customer's car when parking it outside the workshop
Unit A7–L Routinely service the vehicle to maintain optimum performance	You must be able to carry out all types of service as specified by the vehicle manufacturers	Carry out a full service.
Unit A8–L3 Identify faults which affect system performance	To be able to carry out tests and use fault-finding procedures to find out why the vehicle is not performing as it should	Use diagnostic equipment to diagnose why the engine misfires at high speed
Unit A9–L3 Rectify faults in vehicle systems	Once a fault has been diagnosed as above, you must be able to fix it!	Again as above, once the high speed misfire is diagnosed as, say, the rotor arm tracking, you must be able to replace it (not just say it needs a new distributor)!
Unit A10–LH Augment vehicle systems to meet customer requirements	'Augment' means to do work which improves on the original vehicle	Fitting alloy wheels and spotlights
Unit A6–G Reinstate the cleanliness of the vehicle	To be able to valet a car	A complete wash and wax polish and the interior upholstery cleaned

The following units are required for qualification:

Vehicle mechanical and electronic systems Unit replacement (light vehicle level 2)	Vehicle mechanical and electronic systems maintenance and repair (light vehicle level 3)
A1–G	A7–L
A8–L2	A8–L3
A9–L2	A9–L3
A11–L	A10–LH
A12–G	A11–L
A13–G	A12–G
Additional unit A7–L (optional)	A13–G
	Additional unit A1–G (optional)
	Additional unit A6–G (optional)

Collecting evidence for your portfolio

KEY WORDS

■ Auditable evidence
■ Assessments
■ UPK

Because no national examinations are set for an NVQ, you have to collect evidence which shows you can meet the standards. This will usually be collected in a ring binder or similar. The evidence has to be cross-referenced to show that it covers all the criteria in the standards. Some performance criteria may require more than one piece of evidence. When you register for your NVQ, the awarding body (where your certificate will come from, such as City & Guilds) will supply lists of the standards and 'tick boxes' to help you. Your assessor will give lots of help and guidance but you can make a start by collecting evidence from both work and college. Don't forget that some of the evidence from level 2 may also be used at level 3, if it is up to the required standard.

'Evidence' can come from many sources, for example:

- job cards showing work you have done in a garage
- tasks you have carried out while an assessor is observing
- results of tests or examinations
- answers to verbal questions
- written statements made by an expert witness, for example your employer (witness testimony)
- previous certificates
- photographs
- examples of your work
- videos.

In most cases, the types of evidence you supply will be those towards the top of this list. However, if the evidence is good and can be proved, any type can be used. We all have different needs and situations. Your assessor will advise you on what is best in your case. It is also important to note that evidence should come from a task completed after training has taken place, not a training exercise itself. One important point to remember about the evidence you supply is that it should be 'auditable'. This means that, if necessary, your assessor or the verifier could check up on its authenticity. Let me put this in easier words: 'don't try to cheat'!

How to use this book

The previous section talked about 'evidence'. This meant evidence that you can meet the NVQ standards, by being able to do the work and also show an understanding of vehicle technology and workplace systems. This book will help with your understanding, or underpinning knowledge (UPK).

Figure 1.4 Completing portfolio evidence

An important thing to understand about an NVQ is that it is not a course at a training centre or college; it is a qualification by assessment. The assessment will in many cases follow your learning programme.

One way to learn from a book is to read it all the way through! But by the time you get to the end, you've probably forgotten what was at the beginning. To really learn you need to study each part slowly and in detail. You will find that most of the chapters do not relate directly to a particular NVQ unit. This is because an understanding of the vehicle in particular and the motor trade in general is required for almost all of the units.

If you flick through the book you will notice that each chapter (including this one), is generally laid out in the same way. Each chapter starts with an introduction (Start here!) and then covers terminology, the words and abbreviations you will need to know. The next parts deal with the subject in more depth. Towards the end of each main section the key words are highlighted. These are then followed by learning tasks related to the same section.

The idea is that after reading each main section and carrying out the learning tasks, you will have a better understanding of the subject you have just read about. At the very end of each chapter you will again find key words and more learning tasks. UPK statements from the standards are listed at the front of the book together with references to the chapters and sections. These statements will help you to see which areas you have covered. As you complete the learning tasks, you will be able to generate some evidence of UPK for your portfolio. If you do not understand specific areas, you can study the chapter again. It is quite normal to study some chapters more than once.

Good luck with your studies!

LEARNING TASKS

➡ Look back at the key words. Explain each one to a friend, and/or write out a short description to keep as evidence.

➡ Check that you understand what you need to do to gain an NVQ.

I   A modern apprenticeship

What does it mean?   The modern apprenticeship was an idea from the government. Its aim is to formalise apprentice training in the context of NVQs. Just as NVQs try to maintain a consistent approach to different qualifications, so the modern apprenticeships try to maintain a consistent approach between apprenticeships across industry.

Because of the technical developments related to vehicles and their systems, a thorough understanding of principles and concepts is required, as well as practical competences. Key skills are included as part of the overall qualification. Together this ensures the qualification contains skills which are transferable over a wider range of jobs.

Requirements   *NVQ Units*

Unit A1–G	Contract with customers to provide for their vehicle needs
Unit A8–L2	Identify faulty components/units which affect system performance

Unit A9–L2	Remove and replace vehicle components/units
Unit A11–L	Prepare new, used or repaired vehicles for customer use
Unit A12–G	Maintain effective working relationships
Unit A13–G	Maintain the health, safety and security of the working environment
Unit A7–L	Routinely service the vehicle to maintain optimum performance
Unit A8–L3	Identify faults which affect system performance
Unit A9–L3	Rectify faults in vehicle systems
Unit A10–LH	Augment vehicle systems to meet customer requirements
Unit A6–G	Reinstate the cleanliness of the vehicle
C25	Facilitate individual learning through coaching

Key Skills

Communication	Level 2
Information technology	Level 1
Application of number	Level 2
Working with others	Level 3
Improving own learning and performance	Level 3
Problem solving	Level 3

Key skills are the *'key skills'* you will need to carry out your job to a good standard. One of the most important aspects of these skills is that they are

Figure 1.5 We all have to start somewhere!

transferable. In other words, they are useful skills to have for life in general, as well as your work in the motor trade. I know it is hard at times to think ahead, but try to consider the number of opportunities that will be available to you as your career and life progress if you are able to:

■ communicate well
■ use information technology
■ apply and use numbers
■ work with others
■ improve your own learning and performance
■ **and** solve problems.

Put in the work now and it *will* pay off in the future!

Associated underpinning knowledge (UPK)

Schemes of underpinning knowledge are available for modern apprenticeships. Each of these areas will be examined in different parts of the level 2 and 3 books. The 'Light Vehicle Maintenance and Repair' qualification has the following contents:

Engine mechanical	Electric/electronic systems
Cooling system	Vehicle augmentation
Engine lubrication system	Vehicle valeting, preparation and inspection
Fuel system	Customer relations and contracts
Ignition system	Safety
Clutch/fluid couplings	Measurement
Manual transmission	Materials
Automatic transmission	Fixing/securing devices
Final drive/differential	Friction, lubrication, bearings
Driveshafts	Electrics/electronics
Wheels, tyres, wheel bearings	Science
Braking system	Fault diagnosis
Suspension system	Interpreting information
Steering system	Coaching individuals
Body	Key skills

You will be able to cover the above requirements from this and the level 2 book. Please note, though: you will have to access information from a *variety of sources* to cover all the requirements to the necessary standard.

The contract/ agreement

This modern apprenticeship scheme requires all the NVQ units to be achieved as well as a number of key skills. It is designed to formalise apprentice training to NVQ standards and to include key skills. A contract or agreement is required between the local Training and Enterprise Council (TEC), the employer and the employee. This is to ensure that each of the three parties get what they want. Refer to chapter 19 of the level 2 book for further details about contracts. The following table shows the requirements in a simplified form.

Figure 1.6 Your UPK must cover all vehicle systems

Interested party	Requirements/duties
Training and Enterprise Council (TEC)	To pay for the training and ensure quality standard of the programme. Also ensure the national guidelines are followed. If the company can no longer support the apprentice then the TEC will do its best to find another suitable opportunity.
The company	To pay the trainee and provide suitable opportunities and experience for training and development. To keep records and take tests as required.
The apprentice (you?)	To work for the company and observe the company rules. To attend work and training as required (on time) and behave in a responsible manner.

KEY WORDS

- All words in the table
- Modern apprenticeship agreement
- Key skills

A typical contract or agreement will be similar to the following:

This agreement is between	[apprentice's name]
and	[parent/guardian's name (if apprentice under 18)]
and	[company's name]
and	[TEC name]
is made on	[date]

The terms of the agreement are now listed (similar to the above table but with more detail):

Signed	[company]
Date	
Signed	[apprentice]
Date	
Signed	[parent/guardian]
Date	
Signed	[TEC]
Date	

If you enter into a modern apprenticeship agreement, you must be aware of all the details of what you are signing, and then be prepared to carry them out! If this is the best scheme for you, go for it and work hard.

A national traineeship scheme is also available. It is similar to the modern apprenticeship except that the NVQ requirement is at level two.

LEARNING TASKS

➡ Look back at the key words. Explain each one to a friend, and/or write out a short description to keep as evidence.

➡ Write a short explanation about what a contract or agreement means to all parties involved.

2 Health and safety

I Staying healthy and safe

Start here! All of the essential health and safety issues were covered in the level 2 book. However, as you become more experienced, you will have to take more responsibility for this very serious subject. Generally, the key issues remain the same; as you will recall the 'Health and Safety at Work Act' (HASAW) can be split into two areas. Employers have a duty to safeguard the health, safety and welfare of their employees. In other words they must provide:

- A safe place of work with safe access and exit points
- A safe working environment and appropriate welfare arrangements
- Safe systems of work
- Safe plant, equipment and tools
- Safe methods of storing, handling and moving goods
- A procedure for reporting accidents and an accident book
- A safety policy
- Information, instruction, training and supervision where appropriate.

Employees have a duty to safeguard themselves and their workmates. They must:

- Co-operate with their employer to comply with the HASAW
- Take care of their own health and safety as well as that of a workmate
- Not interfere with or misuse any health and safety equipment

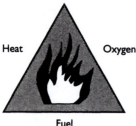

Heat · Oxygen · Fuel

Figure 2.1 Your employer should have a procedure for reporting accidents

Figure 2.2 You should know what to do in an emergency

The HASAW Act places responsibilities on the employer and employee.

As you become more experienced, you will be expected to continually improve your own safe working practices. You will also be expected to take an active part in your employers' duties.

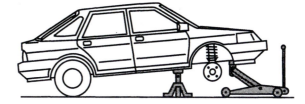

Figure 2.3 You must know how to work safely

Figure 2.4 Warning signs play an important part in driving and in working safely

Terminology	
HASAW	Health and Safety at Work Act. The Act is designed to make us all aware of workplace dangers
COSHH	Control of Substances Hazardous to Health. This is to ensure that an individual or a company takes resposibility for hazardous substances such as old engine oil or cleaning fluids
EPA	Environmental Protection Act. This Act puts very tight controls on the way a business is allowed to affect the environment
Health and Safety Inspectorate	A government agency who make sure the HASAW is followed
Health and safety audit	An inspection carried out in a workshop, for example, to check for anything which may not comply with regulations
Risk assessment	The risks in using a piece of equipment should be assessed and then reduced as much as is reasonably possible
PPE	Personal protective equipment
Responsible person or a Responsible officer	A named person within a company who takes responsibility for health and safety issues, such as keeping the accident book up to date or ensuring proper warning signs are used. Be aware that as your career develops this could soon be you!

LEARNING TASKS

➡ Look back at the key words. Explain each one to a friend, and/or write out a short description to keep as evidence.

➡ Consider how you would provide safety information, instruction, training and supervision in your company or college.

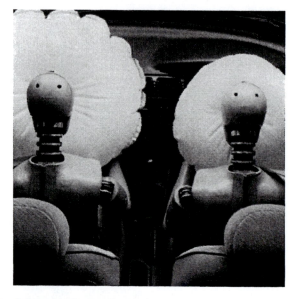

Figure 2.5 Don't be a crash test dummy!

2   Case Study

A company's health and safety policy

The following is an *extract* of a health and safety policy. It is similar to that used by many larger companies but please note I have simplified some areas for reasons of space. It will help you to understand just how far H&S issues reach.

In the discharge of its duty the company will:

1. make itself familiar with the requirements of the Health and Safety at Work, etc Act 1974 and any other health and safety legislation and codes of practices which are relevant to the work of the company, in particular the Management of Health and Safety at Work Regulations 1992

2. ensure that there is an effective and enforceable policy for the provision of health and safety throughout the company

3. periodically assess the effectiveness of this policy and ensure that any necessary changes are made

4. identify and evaluate all risks relating to: accidents, health, company-sponsored activities

5. identify and evaluate risk control measures in order to select the most appropriate means of minimising risk to staff, customers and others.

In particular the company undertakes to provide:

1. a safe place for all employees and customers to work including safe means of entry and exit

2. plant, equipment and systems of work which are safe

3. safe arrangements for the handling, storage and transport of articles and substances

4. safe and healthy working conditions

5. supervision, training and instruction so that all employees can perform their company related activities in a healthy and safe manner. All employees will be offered the opportunity to receive health and safety training which is appropriate to their duties and responsibilities and which will be given before an employee commences any relevant work. Wherever training is required by statute or considered necessary for the safety of staff, customers and others then the company will ensure, within the financial resources available, that such training is provided. All training will be regularly updated

6. necessary safety and protective equipment and clothing together with any necessary guidance, instruction and supervision.

The appointed safety officer will:

1. be aware of the basic requirements of the Health and Safety at Work, etc Act 1974 and any other health and safety legislation and codes of practices relevant to the work of the company

2. ensure, at all times, the health, safety and welfare of staff, customers and others using the company premises or facilities or services or attending or taking part in company-sponsored activities

3. ensure safe working conditions for the health, safety and welfare of staff, customers and others using the company premises and facilities

4. ensure safe working practices and procedures throughout the company including those relating to the provision and use of machinery and other apparatus, so that each task is carried out to the required standards and so that all risks are controlled

5. consult with members of staff, including the safety representatives, on health and safety issues

6. arrange systems of risk assessment to allow the prompt identification of potential hazards

7. carry out periodic reviews and safety audits on the findings of the risk assessment

8. identify the training needs of employees and ensure, within the financial resources available, that all members of employees and customers who have identified training needs receive adequate and appropriate training and instruction in health and safety matters

9. encourage employees to promote health and safety, ensure that any defects in the premises, its plant, equipment or facilities which relate to or may affect the health and safety of staff, customers and others are made safe without delay

10. collate accident and incident information and, where necessary, carry out accident and incident investigations

11. monitor the standard of health and safety throughout the company, including all company-based activities, encourage employees to achieve the highest possible standards and discipline those who consistently fail to consider their own well-being or the health and safety of others

12. monitor first aid and welfare provision.

Supervisory staff will ensure that: (Note that this could quite soon be you!)

1. safe methods of working exist and are implemented throughout their department

2. health and safety regulations, rules, procedures and codes of practice are being applied effectively

3. employees under their jurisdiction are instructed in safe working practices

4. new employees are given instruction in safe working practices

5. regular safety inspections are made of their area of responsibility as required by the appointed safety officer or as necessary

6. positive, corrective action is taken where necessary to ensure the health and safety of all staff, customers and others

7. all plant, machinery and equipment in the department in which they work is adequately guarded

8. all plant, machinery and equipment in the department in which they work is in good and safe working order

9. all reasonably practicable steps are taken to prevent the unauthorised or improper use of all plant, machinery and equipment in the department in which they work

10. appropriate protective clothing and equipment, first aid, and fire appliances are provided and readily available

11. toxic, hazardous and highly flammable substances are correctly used, sorted and labelled

12. they monitor the standard of health and safety, encourage staff, customers and others to achieve the highest possible standards of health and safety and discipline those who consistently fail to consider their own well-being or the health and safety of others

13. all the signs used meet the statutory requirements

14. all health and safety information is communicated to the relevant persons

15. they report, as appropriate, any health and safety concerns to the appropriate individual.

The duties of all employees is to:

1. be familiar with the safety policy and any and all safety regulations laid down by the company

2. ensure health and safety regulations, rules, routines and procedures are being applied effectively by both employees and customers

3. see that all plant, machinery and equipment is adequately guarded

4. see that all plant, machinery and equipment is in good and safe working order

5. not make unauthorised or improper use of plant, machinery and equipment

6. use the correct equipment and tools for the job and any protective equipment or safety devices which may be supplied

7. ensure that toxic, hazardous and highly flammable substances are correctly used, stored and labelled

8. report any defects in the premises, plant, equipment and facilities which they observe

9. take an active interest in promoting health and safety and suggest ways of reducing risks.

Risk assessment

The appointed safety officer will ensure that a risk assessment survey of the premises, methods of work and all company-sponsored activities is conducted annually (or more frequently, if necessary). This survey will identify all defects and deficiencies, together with the necessary remedial action or risk control measures. The results of all such surveys will be reported to the company.

Emergency plans

The appointed safety officer will ensure that an emergency plan is prepared to cover all foreseeable major incidents which could put at risk the occupants or users of the company. This plan will indicate the actions to be taken in the event of a major incident so that everything possible is done to:

1. save time

2. prevent injury

3. minimise loss.

This sequence will determine the priorities of the emergency plan.

The plan will be agreed by the company and be regularly rehearsed by employees and customers. The result of all such rehearsals will form part of the regular risk assessment survey and the outcome will be reported to the company.

The arrangements for first aid provision will be adequate to cope with all foreseeable major incidents. The number of certificated first aiders will not, at any time, be less than the number required by law.

Review

The company will review this policy statement annually and update, modify or amend it, as it considers necessary to ensure the health, safety and welfare of employees and customers.

Now read the following case studies and see if you can make appropriate comments, bearing in mind the health and safety policy you have just read.

The case studies   Good grief, that last section makes heavy reading doesn't it! Note, however, the three key duties of an employer:

- to provide competent workmates
- to provide safe equipment and plant
- to provide a safe system of working.

What can we learn from all that? Read the brief outlines of accidents at work in this section and consider them in the context of the health and safety policy on pages 14–16. As you read each one, bear in mind these two questions:

- Who is to blame?
- How can it be prevented from happening again?

Remember, you can look back to the level 2 book for further information relating to health and safety. The case studies here are not directly related to the motor industry, but I have chosen them because they are real cases and I can give the genuine answers. In most cases you will be able to make a direct comparison to the motor trade. Many of the case studies listed are quite old, but these are often used to set what is called a precedent in law. This means they have a result that other cases can be based on.

1. Mr Cripps was an electrician. His employers sent him, with an apprentice, to install electric lights in a barn. In the barn were some calves, so Mr Cripps told the apprentice to stand at the foot of the ladder and keep the calves away. A calf bumped into the apprentice, who fell against the ladder, knocking it over. Mr Cripps injured his wrist. He sued his employers. (Cripps 1974)

2. Mr Hudson was injured when a practical joke went wrong. The person who played the joke on him had been employed for some four years, and over that time he had often tripped people up and engaged in horseplay, despite having been told not to by his employer. Mr Hudson sued the employer, arguing that the employer had failed to provide a safe and competent workmate. (Hudson 1957)

3. A bus conductor got into a quarrel with a passenger. The bus conductor assaulted the passenger, injuring him. The passenger then sued the bus company, arguing that it was responsible for the acts of its employees. (Keppel Bus Co. 1974)

4. Mrs Woodward was given a lift to work from a colleague. They parked in the factory car park, close to the footpath leading to the factory entrance. She slipped on ice and injured her back. She claimed damages from her employers, arguing that there had been a breach of the factories act. (Woodward 1980)

5. Mr Upson was lifting a metal plate with four workmates. One of the others let go, and the plate slipped, crushing Mr Upson's fingers. He sued his employers, arguing that the system of work was unsafe. (Upson 1975)

6. Mr Litherland worked for a furniture manufacturer, where he used a glue which was often a cause of dermatitis. His employers did not warn him of this risk, nor did they provide him with gloves or barrier cream to reduce his chances of catching dermatitis. Mr Litherland contracted dermatitis and sued his employers. (Litherland 1974)

7. Mr Payne was a labourer in a quarry. One day he was asked to help out by doing the more skilled job of repairing a chain with a steel hammer. He was not supplied with, or told to wear, goggles, nor was he told that if he struck the link pin as opposed to the link, it might shatter. Mr Payne hit the link pin and lost an eye when it shattered. He sued his employers arguing that they had not provided a safe system of work. (Payne 1973)

Figure 2.6 So who is to blame? Not this manufacturer

Before you read the actual results in the next section, see if you can decide on the result for yourself.

The lessons to be learnt

In each case I will give the judgement of the courts (that is, who was to blame or was liable) and then I will suggest ways of preventing the accident happening again.

1. Cripps 1974
Who was to blame?
His employers were liable because the apprentice had been negligent.

How can it be prevented from happening again?
They were responsible for his negligence. It is important to know how much responsibility can be placed on an apprentice. You must note that, as you become more experienced, your role changes from being supervised to supervising.

2. Hudson 1957
Who was to blame?
His employer was liable.

How can it be prevented from happening again?
Clearly, in this case, the employer should have gone further that just warning the 'practical joker', perhaps he should have been sacked?

3. Keppel Bus Co. 1974
Who was to blame?
The bus conductor had not been acting in the course of his employment by striking the passenger and so the company was NOT liable.

How can it be prevented from happening again?
This is very difficult because we don't know all the facts. However, it does raise the issue of how to deal with customers!

4. *Woodward 1980*
Who was to blame?
Her employers were liable. They should have gritted the footpath.

How can it be prevented from happening again?
Floors, passages and stairs must always be kept clean and free of slippery surfaces.

5. *Upson 1975*
Who was to blame?
The employers were liable. They should have arranged (through their foreman) for someone to act as a co-ordinator for the lifting operation.

How can it be prevented from happening again?
Ensure someone takes responsibility for workshop operations.

6. *Litherland 1974*
Who was to blame?
His employers had negligently failed to provide a safe system of work and so they were liable.

How can it be prevented from happening again?
Suitable PPE (personal protective equipment) must be available at all times. Employees must also be made aware of any possible hazards concerning the substances and materials they are dealing with.

7. *Payne 1973*
Who was to blame?
His employers had not provided a safe system of work and so they were liable in damages.

How can it be prevented from happening again?
A safe system of work requires that untrained workers must be told how to look after themselves.

Remember to learn from these incidents, not to copy them!

LEARNING TASKS

➡ Look back at the key words. Explain each one to a friend, and/or write out a short description to keep as evidence.

➡ Examine the case studies as instructed in the main text.

➡ Look at your own workshop or the college workshops and make a note of how you could reduce the risks of accidents. Consider the layout and also the working methods of your workshop.

3 Tools and equipment

3 Introduction

Start here! By now you will be quite competent at using hand tools and basic test equipment. This chapter will therefore concentrate on the more complicated items you may use in the workshop. Diagnostic techniques are very much linked to the use of test equipment. In other words, you must be able to interpret the results of tests. In most cases this involves comparing the result of a test to the reading given in a data book or other source of information.

The oscilloscope is becoming a very popular piece of test equipment. For this reason, I will cover in some detail its operation and use later in the chapter.

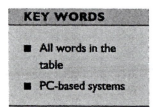

KEY WORDS

- All words in the table
- PC-based systems

Figure 3.1 A selection of air tools

Terminology

Hand tools	Spanners and hammers and screwdrivers and all the other basic bits!
Special tools	A collective term for items not held as part of a normal tool kit. They could also be tools required for just one specific job
Test equipment	In general, this means measuring equipment. Most tests involve measuring something and comparing the result of that measurement to manufacturer's data. Measuring devices can range from a simple ruler to an engine analyser
Dedicated test equipment	Some equipment will only test one specific type of system. The large manufacturers supply equipment designed specifically for their vehicles. For example, a diagnostic device which plugs in to a certain type of fuel injection ECU
Accuracy	Careful and exact, free from mistakes or errors, and adhering closely to a standard
Calibration	Checking the accuracy of a measuring instrument
Serial port	A connection to an electronic control unit, for example, a diagnostic tester or computer. 'Serial' means the information is passed in a 'digital' string like pushing black and white balls through a pipe in a certain order
Code reader or scanner	This device reads the 'black and white balls' mentioned above or the on-off electrical signals, and converts them into information we can understand
Combined diagnostic and information system	Usually PC-based, these systems can be used to carry out tests on vehicle systems, and they also provide an electronic workshop manual. Computer-guided test sequences can also be carried out
Oscilloscope	The main part of an oscilloscope or 'scope' is the display which is like a TV or computer screen. A scope is a voltmeter but instead of readings in numbers it shows the voltage levels by a 'trace' or mark on the screen. The marks on the screen can move and change very fast allowing us to see the way voltages change

Basic tools

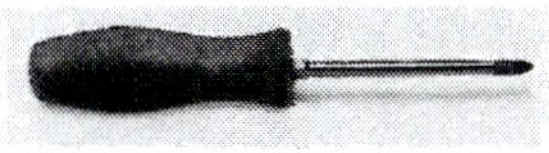

Figure 3.2
A screwdriver!

■ Flat tip
▮ Scrulox®
✚ Phillips®
✸ Torx®
⬢ Hex
▮ Clutch 'G'
✖ Pozidriv®
人 Tri-wing®
▰ Clutch 'A'
✸ Tamper resistant torx®

Figure 3.3 Many types of 'driver' shapes are now in use

You cannot learn to use tools from a book; it is clearly a very practical skill. However, you can follow the recommendations made here and, of course, by the manufacturers. Even the range of basic hand tools is now quite daunting and very expensive. One thing to highlight as an example is the number of different types of screwdriver ends (Figure 3.3). These are worth mentioning because using the wrong driver and damaging the screw head causes a lot of trouble. And, of course, screwdrivers are all available in many different sizes!

It is worth repeating the general advice and instructions for the use of hand tools:

■ Only use a tool for its intended purpose.
■ Always use the correct size tool for the job you are doing.
■ Whenever possible, pull a wrench rather than pushing it.
■ Do not use a file or similar without a handle.
■ Keep all tools clean and replace them in a suitable box or cabinet.
■ Do not use a screwdriver as a pry bar.
■ Always follow manufacturer's recommendations (you cannot remember everything).
■ Look after your tools and they will look after you!

Special tools   This general heading covers all tools which are not considered to be standard equipment. Spanners, sockets and hammers would be seen as standard tools. Most of these were covered briefly in the level 2 book.

The number of special tools available makes it impossible to cover all the details in this section. They range from the tools essential for doing a job, to the ones that make our lives easier. Simple things like a ratchet ring spanner, whilst not essential, certainly make some jobs much easier. I'm sure, like me, you spend time looking at all the tools in a catalogue that you would like but can't afford! An important point to note here though, is that if you spend money on equipment, which means you are able to complete a job in less time, should the customer pay less? This is a difficult one, and is a good reason for charging jobs based on standard times. It's fair to both the technician and the customer.

Using any new special tool requires the usual basic recommendation: refer to the manufacturer's instructions. Rather than try to cover the whole range of special tools, I will use different pullers as examples. The basic procedures will apply to all types of tools. Figure 3.4 shows a selection of 'jaw' type pullers. Snap-on® provides the following guide: the ABC's of selecting the right puller.

A.   What type of puller do I want to use? Will it be the combination that works best?

B.   Determine the 'reach' and 'spread' needed. The puller's reach must equal or exceed the removal distance of the part. The puller's spread is the width of the part (Figure 3.5).

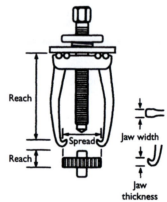

Figure 3.4 Two and three legged pullers

Figure 3.5 Reach and spread of a puller

C.   Estimate the puller's safe working load. Generally, if you have the correct 'reach and spread', you will have the correct safe working load. If in doubt, use the next larger size.

A general rule for manual pullers is that the pressure screw should be at least half the diameter of the shaft on the pulling/pushing job. The particular type of puller will vary depending on the job you are doing. Figure 3.6 a, b and c show three different situations that require some kind of puller.

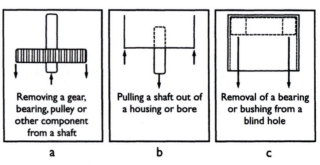

a	b	c
Removing a gear, bearing, pulley or other component from a shaft	Pulling a shaft out of a housing or bore	Removal of a bearing or bushing from a blind hole

Figure 3.6 Different tasks for a puller

a) A 2 or 3 jaw type puller with outside jaw positions is the best choice.

b) A push-pull puller is best, or a slide hammer if the case lacks strength to push against as shown.

c) An inside jaw puller (same as 'a' with the legs reversed) is the best choice. Expanding collets may be required to grip the inside surface of the bearing or bush.

LEARNING TASKS

➡ Look back at the key words. Explain each one to a friend, and/or write out a short description to keep as evidence.

➡ Make a simple sketch to show a puller removing a flywheel.

➡ Make a list of special tools used as part of your work. For each, state two safety points and two advantages.

---

2 Test equipment

Basic test meters

An essential tool for working on vehicle electrical and electronic systems is a good digital multimeter. Digital meters are most suitable for accuracy of reading as well as available facilities. The following list of functions, beginning with the essential then moving down to the desirable, should be considered.

KEY WORDS

- All words in the table
- Accuracy
- Scanner
- Exhaust gases
- Instruments

Function	Range	Accuracy
DC voltage	500 V	0.3%
DC current	10 A	1.0%
Resistance	0 to MΩ	0.5%
AC voltage	500 V	2.5%
AC current	10 A	2.5%
Dwell	3,4,5,6,8 cylinders	2.0%
RPM	10,000 rpm	0.2%
Duty cycle	% on/off	0.2% /kHz
Frequency	over 100 K	0.01%
Temperature	>900 °C	0.3% + 30C
High current clamp	1000 A DC	Depends on conditions
Pressure	3 bar	10.0% of standard scale

A way of determining the quality of a meter, apart from the facilities provided, is to consider the following:

- accuracy
- loading effect of the meter
- protection circuits.

The loading effect is a consideration for any form of measurement. The question to ask is, 'Does the instrument change the conditions so my reading is incorrect?' With a multimeter, this relates to the internal resistance of the meter. It is recommended that the internal resistance of a meter should be a minimum of 10 MΩ. This not only ensures greater accuracy, but also prevents the meter damaging sensitive circuits.

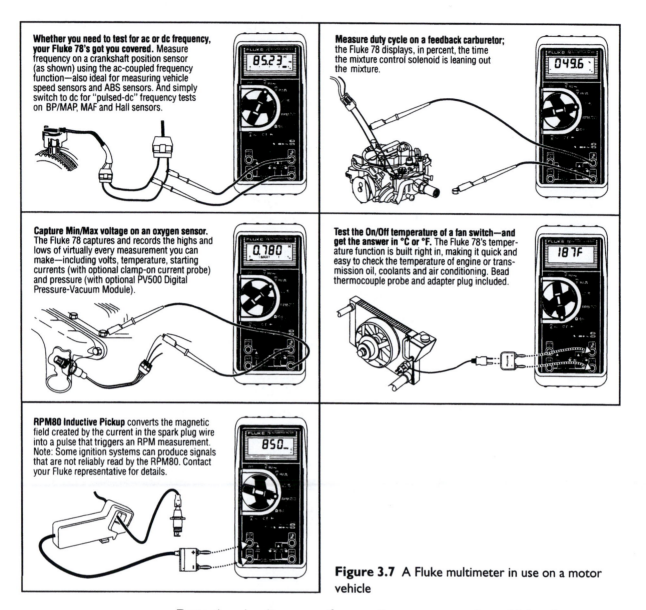

Whether you need to test for ac or dc frequency, your Fluke 78's got you covered. Measure frequency on a crankshaft position sensor (as shown) using the ac-coupled frequency function—also ideal for measuring vehicle speed sensors and ABS sensors. And simply switch to dc for "pulsed-dc" frequency tests on BP/MAP, MAF and Hall sensors.

Measure duty cycle on a feedback carburetor; the Fluke 78 displays, in percent, the time the mixture control solenoid is leaning out the mixture.

Capture Min/Max voltage on an oxygen sensor. The Fluke 78 captures and records the highs and lows of virtually every measurement you can make—including volts, temperature, starting currents (with optional clamp-on current probe) and pressure (with optional PV500 Digital Pressure-Vacuum Module).

Test the On/Off temperature of a fan switch—and get the answer in °C or °F. The Fluke 78's temperature function is built right in, making it quick and easy to check the temperature of engine or transmission oil, coolants and air conditioning. Bead thermocouple probe and adapter plug included.

RPM80 Inductive Pickup converts the magnetic field created by the current in the spark plug wire into a pulse that triggers an RPM measurement. Note: Some ignition systems can produce signals that are not reliably read by the RPM80. Contact your Fluke representative for details.

Figure 3.7 A Fluke multimeter in use on a motor vehicle

Protection circuits are worth a mention, as many motor vehicle voltage readings are prone to high voltage transient spikes that can damage low quality equipment. A fused current range is also to be recommended. Figure 3.7 shows a multimeter in use on a motor vehicle.

Pressure gauge   Measuring the fuel pressure on a fuel injection engine is of great value when fault finding. Many types of pressure gauge are available and often come as part of a kit consisting of various adapters and connections. The principle of the gauges is that they contain a very small tube wound in a spiral. As fuel under pressure is forced into a spiral tube, it unwinds causing the needle to move over a graduated scale. Figure 3.8 shows a pressure gauge kit, and a gauge connected to a vehicle.

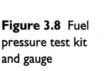

Figure 3.8 Fuel pressure test kit and gauge

Engine analysers   Some form of engine analyser has become an essential tool for fault finding modern vehicle engine systems. Many of the latest machines are based around a personal computer. This allows more facilities, which can be added to by simply changing the software. Whilst engine analysers are designed to work specifically with a motor vehicle, it is worth remembering that the machine consists basically of three parts:

- multimeter
- gas analyser
- oscilloscope.

This is not intended to imply that other available tests, such as cylinder balance, are less important, but to show that the analyser is well suited to presenting results of electrical tests in a convenient way to allow diagnosis of faults.

The key component of any engine analyser is the oscilloscope which allows the user to 'see' the signal under test. Two types of oscilloscope are available: analogue or digital. Figure 3.9 shows the basic principal of an analogue oscilloscope. Heating a wire creates a source of electrons. The electrons emitted are accelerated by suitable voltages and focused into a beam. This beam is directed towards a fluorescent screen where it causes light to be given off. The plates as shown in Figure 3.9 are known as 'x' and 'y' plates as they make the electron beam draw a 'graph' of a voltage signal. The 'x' plates are supplied with a signal, which causes the beam to move across the screen from left to right and then to 'fly back' and start again (this is known as the time base). The beam moves because it is attracted towards whichever plate has a positive voltage. The 'y' plates can now be used to show voltage variations of the signal under test. The time base can be adjusted automatically, as is the case with many analysers, or manually on a stand-alone oscilloscope. The signal from the item under test can either be amplified or reduced, much like changing the scale on a voltmeter. The trigger – in other words, when the trace across the screen starts – can be caused internally or externally. In the case of the engine analyser, triggering is often external, whenever an individual spark fires or each time number one spark plug fires.

A digital oscilloscope has much the same end result as the analogue type, but the signal can be thought of as being plotted rather than drawn on the screen. The test signal is converted from a voltage to a digital number and the time base is an electronic timer circuit. Because the signal is plotted on a screen from data in memory, the 'picture' can be saved, frozen or even printed. This technique is becoming the norm as the display can be made easier to read by including scales and notes or showing two or more traces for comparison. Figure 3.10 shows the principle of a digital oscilloscope.

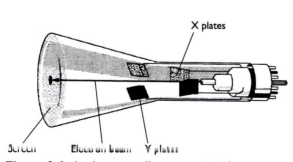

Figure 3.9 Analogue oscilloscope principle

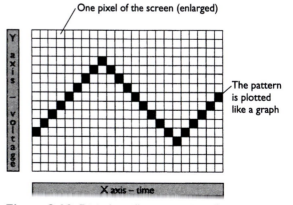

Figure 3.10 Digital oscilloscope principle

The modern trend with engine analysers seems to be to allow both guided test procedures with pass/fail recommendations for the less skilled technician, and freedom to test any electrical device using the facilities available in any reasonable way. This is more appropriate for the highly skilled technician. Some of the routines available on modern engine analysers are listed below:

- tune-up
- symptom analysis
- waveforms
- adjustments
- MOT emissions test.

The connections to the vehicle for standard use are much the same for most equipment manufacturers. These are listed in the following table.

Connection	Purpose or example of use
Battery positive	Battery and charging voltages
Battery negative	A common earth connection
Coil positive	To check supply voltage to coil
Coil negative (adapters are available for DIS) primary waveforms	To look at dwell, rev/min and
Coil HT lead clamp (adapters are available for DIS)	Secondary waveforms
Number one cylinder plug lead clamp	Timing light and sequence of waveforms
Battery cable amp clamp	Charging and starting current
Oil temperature probe (dip stick hole)	Oil temperature
Vacuum connection	Engine load

Figure 3.11 Snap-on Counselor

Figure 3.11 shows the Snap-on Counselor digital oscilloscope engine analyser. This test equipment has the following features:

- high tech test screens. Vacuum waveform, cylinder time balance bar graph, power balance waveform and dual trace lab scope waveform
- scanner interface. This allows the technician to observe all related information at the same time
- auxiliary display screen capability. The Counselor can use a remote monitor if it has suitable connectors
- simplified keyboard design. Keys are organised in logical groupings and colour-coded for easy recognition
- expanded memory. This feature allows thirty two screens to be saved at once, which can be recalled at a later time for evaluation and reference
- multiple power selection. In addition to mains supply, the Counselor can be operated from a 12V battery. Its compact size and 12V option allows it to be used on test drives
- portability. The Counselor can be easily moved from one bay to another rather than tying up one bay for a specific purpose.

Exhaust gas measurement It has now become standard to measure four of the main exhaust gases, namely:

- carbon monoxide (CO)
- carbon dioxide (CO_2)

- hydrocarbons (HC)
- oxygen (O_2).

The emission test module is often self contained with its own display, but it can be linked to the main analyser display. Often lambda value and the air fuel ratio are displayed in addition to the four gasses. The Greek symbol lambda (λ) is used to represent the ideal air fuel ratio (AFR) of 14.7:1 by mass. In other words, this is just the right amount of air to burn up all the fuel. Typical gas, lambda and AFR readings are given for a closed loop lambda control system, before (or without) and after the catalytic converter. The figures below are for a modern engine in excellent condition (these are only examples – you should always check current data).

	CO%	HC_{ppm}	CO_2%	O_2%	Lambda (λ)	AFR
Before catalyst	0.6	120	14.7	0.7	1.0	14.7
After catalyst	0.2	12	15.3	0.1	1.0	14.7

The composition of exhaust gas is now a critical measurement and a particular degree of accuracy is required. To this end the infrared measurement technique has become the most suitable for CO, CO_2 and HC. Oxygen is measured by electro-chemical means in much the same way as the on-vehicle lambda sensor. Figure 3.12 shows the Snap-on four-gas emission analyser. This has the following features:

- stand alone unit is not dependant on other equipment
- graph screens simultaneously display up to four values as graphs, and the graph display order is user-selectable. Select from HC, CO, CO_2, O_2 and rev/min for graph display
- user can create personalised letterhead for screen printouts
- uses the non-dispersive infrared (NDIR) method of detection (each individual gas absorbs infrared light at a specific rate)
- display screens may be 'frozen' or stored in memory for future retrieval
- recalibrates at the touch of a button (if calibration gas and a regulator are used)
- displays exhaust gas concentrations in real time numerics, or create live exhaust gas data graphs in selectable ranges
- calculates and displays lambda (λ) (the ideal air fuel ratio of about 14.7:1)
- displays engine rev/min in numeric or graph form and display oil temperature along with current time and date
- displays engine diagnostic data from a scanner
- operates from mains supply or a 12V battery.

Accurate measurement of exhaust gas is not only required for MOT tests but is essential to ensure an engine is correctly tuned. Figure 3.13 shows the make up of a typical exhaust. Note the toxic emissions are in fact very small, but none-the-less very dangerous.

Figure 3.12 Snap-on exhaust gas analyser

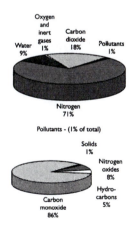

Figure 3.13 Composition of exhaust

Dedicated test equipment

As the electronic system of modern vehicles becomes more complicated, developments in suitable test equipment must follow. The term 'dedicated' implies test equipment used for only one specific system. Figure 3.14 is a representation of one type of dedicated test equipment. A special plug and socket is used to 'break in' to the ECU wiring, whilst generally still allowing the vehicle system to function normally.

Figure 3.14 Lucas Laser 2000 tester

Readings can be taken between various points and compared to set values, allowing diagnosis. Ford have used a system such as this for many years, simply known as a breakout box. A multimeter takes the readings between predetermined test points on the box, which are connected to the ECU wiring.

A further development of this system is a digitally-controlled tester that will run very quickly through a series of tests and display the results. These can be compared with stored data allowing a pass/fail output.

Many electronic systems now have ECU's that contain self-diagnosis circuits. This is represented by Figure 3.15. Activating the blink code output can access the information held in the ECU memory. This is done in some cases by connecting two wires and then switching on the ignition. A further refinement is to read the information via a serial link, which requires suitable test equipment.

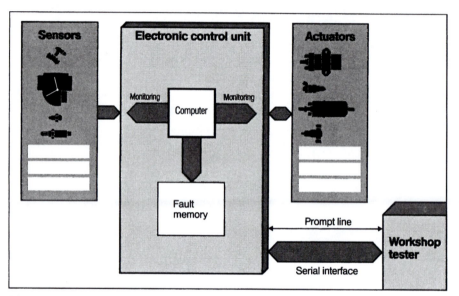

Figure 3.15 Engine control with self-diagnosis

Serial port communications – the scanner

Serial communication is an area that is continuing to grow. A special interface is required to read data. This standard is designed to work with a single or two wire port which connects vehicle electronic systems to a diagnostic plug. Many functions are then possible when a scanner is connected.

Possible functions include the following:

- Identification of ECU and system to ensure the test data is appropriate to the system currently under investigation.
- Read out of current live values from sensors so spurious figures can be easily recognised. Information such as engine speed, temperature air flow and so on can be displayed and checked against test data.
- System function stimulation allows actuators to be tested by moving them and watching for suitable responses.
- Programming of system changes. Basic idle CO or changes in basic timing can be programmed into the system.

At present, several standards exist which means several different types of serial readers are needed – or, at best, several different adapters and program modules. A new standard called On Board Diagnostics II (OBD II) has been developed by the Society of Automotive Engineers (SAE). In the USA, all

Figure 3.16 GenRad scanner

new vehicles must conform to this standard. This means that just one scan tool will work with all new vehicles. This standard is almost certain to be adopted in the UK.

A company called GenRad has produced a scanner to meet these standards. Figure 3.16 shows the GDS500 scanner. This scanner allows the technician to perform all the necessary operations, such as fault code reading, via a single common connector. The portable hand-held tool has a large graphics display allowing instructions and data to be presented clearly. Context-sensitive help is available to eliminate the need to refer back to manuals to look up fault code definitions. It has a memory, so data can be reused even after disconnecting power from the tool. This scanner will even connect to the Controller Area Network (CAN) systems. This is a very new type of wiring system using a data bus.

Scopemeter

A very good piece of equipment becoming more popular is the 'Scopemeter' (Figure 3.17). This is a hand-held digital oscilloscope, which allows data to be stored and transferred to a PC for further investigation. The Scopemeter can be used for a large number of vehicle tests. The waveforms used as examples at the end of this chapter were 'captured' using a Scopemeter. This type of test equipment is highly recommended.

Figure 3.17 Bosch portable multiscope

New developments in diagnostics

The next development in diagnosis and testing is likely to be by increased networking of vehicles via adapters to computers locally, and then via modems and the telephone lines. It is already quite common practice in the computer industry, using suitable hardware and software, to remotely link one computer to another in order to carry out diagnostics and repairs. This technique can be extended to the computerised systems on modern vehicles. Access to the latest data and test procedures is available at the 'touch of a screen' or the 'click of a mouse'.

The incredible storage capacity of compact disks means they are used more and more as the medium for information storage. When used in conjunction with test equipment, as described earlier, the operator will be able to work through the most complex of faults, with interactive help from the computer.

Figure 3.18 GenRad
diagnostic system

GenRad have a system very similar to the above description. It is called the GDS3000 and is a fully integrated information and diagnostic system. It links faultfinding procedures and electronic workshop service manuals together. The GDS3000 is based on industry standard PC technology using a 486 DX2 or DX4 processor. It will run 'normal' programs as well as the specialist diagnostic and information programs for use in the workshop. Figure 3.18 shows this comprehensive piece of test equipment. Most major manufacturers have similar devices, Rover, for example, use a system known as 'Test Book'.

Micrometer A micrometer is a measuring instrument designed to measure to an accuracy of 0.01 mm. Its principle of operation is quite simple: a very accurately manufactured screw thread is used with a pitch of 0.5 mm. This means that as it is turned one complete turn will move it 0.5 mm. A complex mathematical calculation using your calculator should therefore tell you that two complete turns will move it 1.0 mm! A main scale is marked on the micrometer with 0.5 mm marks so you can see the distance moved. A rotating scale marked from 0 to 50 is used to give the required accuracy.

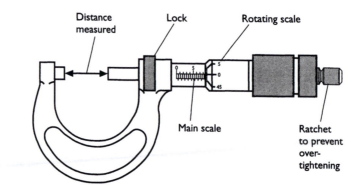

Figure 3.19 A 0–25 mm micrometer

The way to read a micrometer is quite simple. After it has been gently screwed, using the thimble ratchet, across whatever you want to measure – then simply follow the 'ABC':

A. Find the nearest whole millimetre.

B. Add the next half millimetre if showing (0.5).

C. Add the hundredths of a millimetre shown on the rotating scale.

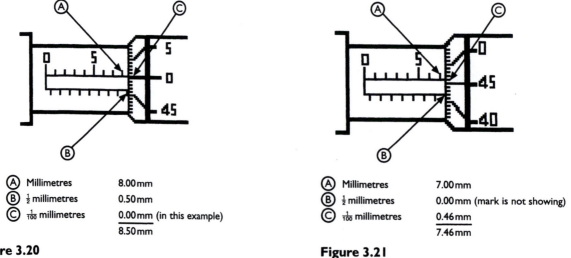

Ⓐ Millimetres	8.00 mm
Ⓑ ½ millimetres	0.50 mm
Ⓒ 1/100 millimetres	0.00 mm (in this example)
	8.50 mm

Figure 3.20

Ⓐ Millimetres	7.00 mm
Ⓑ ½ millimetres	0.00 mm (mark is not showing)
Ⓒ 1/100 millimetres	0.46 mm
	7.46 mm

Figure 3.21

One word of warning, though: micrometers are available in size ranges of 0–25 mm, 25–50 mm, 50–75 mm, and so on. Make sure you remember this when taking a measurement and add 25 mm or 50 mm as appropriate.

The most common uses for a micrometer in the motor trade are for measuring valve shims and measuring crankshaft journals. Often using an instrument such as a micrometer looks difficult. Well it's not; a bit of practice and you will soon find it very easy.

Dial gauge

A dial gauge is a device used to test very accurately the movement of something. A good example is the run out of brake discs. A plunger on the dial gauge is made to run up against the disc as it is turned. The body of the gauge is clamped in position and, via very accurate gears, the movement of the plunger makes a hand turn on a clock dial. The face of the clock is marked off in 0.01 mm increments and will usually rotate enough times to measure up to 10 mm. Diesel pump timing is often set or checked with a dial gauge. The pumping plunger is made to act on the dial gauge so that as the engine is turned it is possible to tell the position of the injection plunger very accurately.

Accuracy of test equipment

'Accurate' can mean a number of slightly different things:

- careful and exact
- free from mistakes or errors
- precise
- adhering closely to a standard.

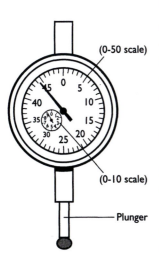

Figure 3.22 Dial gauge

Consider measuring a length of wire with a steel rule. How accurately could you measure it? To the nearest 0.5 mm? This raises a number of issues. Firstly, you could make an error reading the ruler. Secondly, why do we need to know the length of a bit of wire to the nearest 0.5 mm? Thirdly, the ruler may have stretched and so may not give the correct reading!

The first and second of these issues can be avoided by knowing how to read the test equipment correctly, and also by knowing the appropriate level of accuracy required. Would a micrometer be suitable for a plug gap? A ruler for valve clearances? I think you get the idea. The accuracy of the equipment itself is another issue altogether.

'Accuracy' is a term meaning how close the measured value of something is, compared to its actual value. For example, if a length of about 30 cm is measured with an ordinary wooden ruler, then the error may be up to 1 mm too high or too low. This is quoted as an accuracy of ± 1 mm. This may also be given as a percentage, which in this case would be 0.33%.

Resolution or, in other words, the 'fineness' with which a measurement can be made, is related to accuracy. If a steel ruler was made to a very high standard but only had one marking per centimetre, it would have a very low resolution even though the graduations were very accurate. In other words, the equipment is accurate but your reading will not be!

To ensure instruments are, and remain accurate, there are just two simple guidelines:

- look after the equipment; a micrometer thrown on the floor, for example, will not be accurate

    ensure instruments are calibrated regularly – this means being checked against known good equipment.

Here is a summary of the steps to ensure a measurement is accurate:

Step	Example
Decide on the level of accuracy required	Do we need to know that the battery voltage is 12.6 V or 12.635 V?
Choose the correct instrument for the job	A micrometer to measure the thickness of a shim
Ensure the instrument has been looked after and calibrated when necessary	Most instruments will go out of adjustment after a time. You should arrange for adjustment at regular intervals. Most tool suppliers will offer the service, or in some cases you can compare older equipment to new stock
Study the instructions for the instrument in use and take the reading with care. Ask yourself if the reading is roughly what you expected	is the piston diameter 70.75 mm or 170.75 mm?
Make a note if you are taking several readings	Don't take a chance: write it down

LEARNING TASKS

➡ Look back at the key words. Explain each one to a friend, and/or write out a short description to keep as evidence.

➡ Make a list showing the stages involved in taking voltage measurements, torque settings and shim thickness.

➡ Examine a real vehicle and consider the range of test equipment required to test its systems.

➡ Write a short explanation about how to connect an engine analyser to an engine and the purpose of each connection.

3   Waveforms

Introduction

In this section, I will first explain the principle of using an oscilloscope for displaying waveforms and then examine a selection of actual waveforms. You will find that 'waveform' and 'patterns' are used in books and workshop manuals but they mean the same thing.

KEY WORDS

■ Oscilloscope
■ Waveform
■ Time base
■ Volts/div
■ Time/div

When you look at a waveform on a screen, you must remember that the height of the scale represents 'voltage' and the width represents 'time'. Both of these axis can have their scales changed. They are called 'axis' because the 'scope' is drawing a graph of the voltage at the test points over a period of time. The time scale can vary from a few µs to several seconds. The voltage scale can vary from a few mV to several kV. For most test measurements, only two connections are needed just like a voltmeter. The time scale will operate at intervals pre-set by the user. It is also possible to connect a 'trigger' wire so that, for example, the time scale starts moving across the screen each time the ignition coil fires. This keeps the display in time with the speed of the engine. When you use a full engine analyser, all the necessary connections are made as listed in a previous section. Figure 3.23 shows an example of a waveform.

For each of the following waveforms I have noted what is being measured, the time and voltage settings and the main points to examine for correct operation. All the waveforms shown are from a correctly operating vehicle. The skill you will learn by practice is to note when your own measurements vary from those shown here.

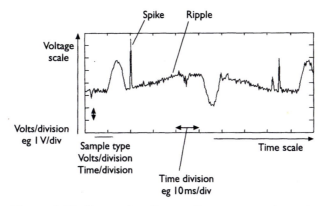

Figure 3.23 How to 'read' an oscilloscope trace (a random signal is shown)

Here is a reminder of some abbreviations that will be useful for the next section (see background studies in the level two book for more details):

- mV   millivolts   (there are 1000 mV in 1 V)
- kV   kilovolts   (1000 V = 1 kV)
- μs   microsecond   (there are 1 000 000 μs in 1 s)
- ms   milliseconds   (there are 1000 ms in 1 s).

Ignition
- Inductive pulse generator output (Figure 3.24).
- Hall effect pulse generator output (Figure 3.25).
- Primary circuit pattern (Figure 3.26).
- Secondary circuit pattern (one cylinder shown as Figure 3.27).
- Secondary circuit pattern (four cylinders shown – called parade as Figure 3.28).

Charging
- Alternator ripple voltage (Figure 3.29).

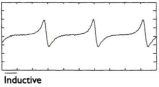

Inductive
1 V/div
10 ms/div

Figure 3.24 A regular smooth output

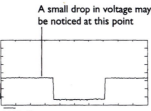

Hall effect
3 V/div
5 ms/div

Figure 3.25 Clean switching from high to low

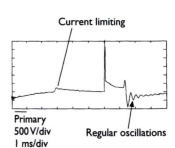

Primary
500 V/div
1 ms/div

Figure 3.26 A regular steady pattern

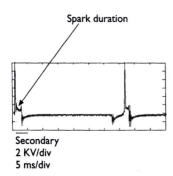

Secondary
2 KV/div
5 ms/div

Figure 3.27 A peak of about 12 kV

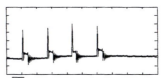

Secondary parade
4 kV/div
1 ms/div

Figure 3.28 All spikes should be within a few kV

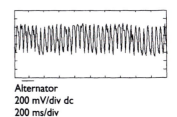

Alternator
200 mV/div dc
200 ms/div

Figure 3.29 A regular pattern and frequency

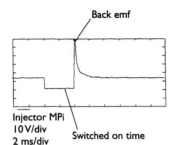

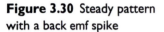

Figure 3.30 Steady pattern with a back emf spike

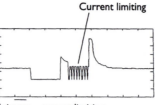

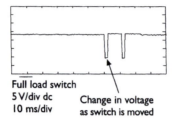

Figure 3.31 Regular current limiting pattern

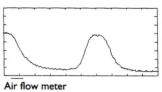

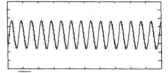

Figure 3.32 Smooth changes as flow changes

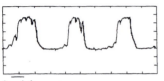

Figure 3.33 Fairly regular oscillations from about 0.2 to 0.8 V (200 to 800 mV)

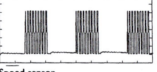

Figure 3.34 Clean switching

Figure 3.35 A regular sine wave

Figure 3.36 An even regular pattern

Injection
- Injector waveform (Figure 3.30).
- Injector waveform with current limiting (Figure 3.31).
- Air flow meter output (flap type Figure 3.32).
- Lambda sensor voltage (Figure 3.33).
- Full load switch operation (Figure 3.34).

General
- ABS wheel speed sensor output signal (Figure 3.35).
- Vehicle speed sensor (Figure 3.36).

LEARNING TASKS

➡ Look back at the key words. Explain each one to a friend, and/or write out a short description to keep as evidence.

➡ Make a simple sketch to show either waveforms you have seen indicating a fault or the results you would expect from various faults. For example, waveforms indicating open circuit plug lead, loose ABS wheel speed sensor and dirt on an air flow sensor variable resistor.

➡ Write a short explanation on the advantages of system testing with a 'scope'.

4 Engines

1 Introduction

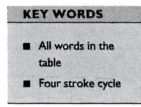
The basic operation of an engine was covered in the first book. This chapter will therefore look in more detail at the materials used and the design of an engine. Multi-valve engines and turbochargers will also be examined. Case studies are presented about well-known engines to introduce and explain common techniques and methods.

Remember, this chapter concentrates on the general technology of engines. You must therefore always, 'Access appropriate data from appropriate sources,' or, look up the information that relates to the particular engine you are working on at the time! Figure 4.1 shows the BMW 325tds turbo diesel engine.

Figure 4.1 The BWM 325tds turbo diesel engine uses advanced technology to produce excellent performance

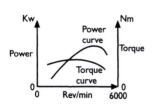

Figure 4.2 Typical modern vehicle torque and power curves

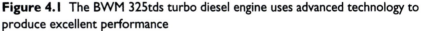

LEARNING TASKS

➡ Look back at the key words. Explain each one to a friend, and/or write out a short description to keep as evidence.

➡ Make a simple sketch to show the drive arrangement of a DOHC engine.

➡ Examine a number of real systems and note the different techniques and layouts used.

➡ Look back to the level 2 book and revise the basic principles of engine operation.

Terminology

DOHC	Double overhead camshaft. In most cases one cam acts on the inlet valves and one on the exhaust valves
Multi-valve engines	This describes any engine with more that two valves per cylinder. The benefits are a larger opening for the entry of new mixture or the outlet of the exhaust. Improving what is known as 'volumetric efficiency'
Wet liner	An engine cylinder that is in direct contact with the coolant. This gives a good cooling effect and also allows a suitable material to make contact with the piston. For example, an aluminium block must be fitted with steel or cast iron liners because, otherwise, it would wear out very quickly
Dry liner	An engine cylinder liner not in direct contact with the coolant
Swirl	'Barrel swirl' is the term used to describe turbulence inside the cylinder, which swirls across the axis of the bore, rather than around it. Turbulence around the axis of the bore is called 'axial swirl'. The swirling or rotary movement is passed on to the mixture entering a cylinder by offsetting the inlet tract. This is done to aid the combustion process
Squish	This describes what happens to the mixture in the combustion chamber as it is compressed. One area, known as a 'squish zone', may be designed in order to improve the way the cylinder charge behaves
Maximum power	This is given in kilowatts (kW), horsepower (HP), or even PS, which is one *metric* horsepower! It is quoted at a certain engine speed. Note: 1 HP = 745.7 W and 1 PS = 735.5 W A typical 1100 cc engine may be rated as: 44 kW @ 6000 rev/min
Maximum torque	This is given in Newton metres (Nm), or in some cases pounds feet (lb.ft). It is quoted at a certain engine speed. Note: 1 lb.ft = 1.35 Nm A typical 1100 cc engine may be rated as: 89 Nm @ 3500 rev/min
Stretch bolts	Bolts used in a significant number of engines to secure the cylinder head. They stretch when tightened to the correct torque. This provides the necessary clamping load of the cylinder head to the block. These bolts must be renewed if the head is removed. Engines such as the Rover 'K' series, discussed later, use long stretch bolts which run from the head to the crankcase
Naturally aspirated	Most types of engines are naturally aspirated. This means the air or air-fuel mixture is drawn into the cylinder on the induction stroke without pressurising. In other words, the engine is not turbocharged or supercharged
Turbocharger	A turbine driven by the exhaust gases drives a compressor, which forces more air or air-fuel charge into the cylinder
Supercharger	This is a compressor similar to a turbocharger, mechanically driven by the engine
Balance shaft	The movement of the pistons and crankshaft in an engine causes vibration. A balance shaft is fitted with counter balance weights rotating in time with the crankshaft, reducing the vibrations which could cause serious damage to the engine's mechanical components
Volumetric efficiency	The volume of air or air-fuel mixture taken into the cylinder when an engine is operating compared to the cylinder's volume. Most naturally aspirated engines work at about 80% efficiency

2   Materials and components

Introduction   There is a large number of different components in a modern engine. The variety of different materials used is also extensive. However, the main components of an engine are common to most designs. The following section lists these components, together with a description of each and a short discussion about the type of materials used.

Figure 4.3 Double overhead camshaft with four valves per cylinder

As your understanding of technology develops, you will realise that in many situations the methods and materials are chosen for a number of reasons. The properties of materials are important, but then so are the costs!

Engine components

Name	Functional requirements	Materials
Piston	The piston must perform a number of functions: 1. Transfer the combustion pressure to the connecting rod and support the resulting force against the cylinder wall 2. Guide the connecting rod in the cylinder 3. Seal the combustion chamber from the crankcase 4. Dissipate its heat energy to the coolant via the piston rings Combustion chamber and piston crown shape determine the way in which forces are transferred via the gudgeon pin to the connecting rod	The most common material for cylinder liners is grey cast iron and aluminium is most commonly used for pistons. Pistons and cylinder liners, therefore, have different rates of expansion. Piston skirt clearance in the cylinder, however, must be kept to a minimum. This is to reduce 'piston slap' and improve sealing. Invar steel strips or similar materials are sometimes cast into the piston to limit its expansion. The piston weight, which should be as low as possible, coupled with the temperatures at which it works, causes a load which is often near the limit of the piston material strength
Piston rings	Piston rings are used to make a gas-tight seal between the combustion chamber and the engine crankcase. The two top rings are gas seals. At least one further specially-shaped ring serves as an oil control ring, and ensures proper oil conditions at the piston and its other rings. Due to the very high, tensioning forces, piston rings are a major source of friction in a piston engine	The piston rings are generally made from cast iron. The top ring can be chromium plated to improve its properties for high performance applications. The main parts of the oil scraper rings are often made from steel, which is chromium plated on the contact surfaces. A number of different types of piston ring are used

Name	Functional requirements	Materials
Connecting rod	The connecting rod connects the piston to the crankshaft. It is subjected to very high compression, tensile and bending stresses. It carries the bearings for the gudgeon pin and the big end. The piston stroke and the crankshaft counterweight radius determine the length of the connecting rod	In many smaller engines, particularly those subject to lower loads, cast iron alloy con-rods are now popular. Con-rods for high performance engines are usually drop forged steel
Crankshaft	Reciprocating piston movements are transferred by the connecting rods to the crankshaft. As a result of the forces, torque, bending moments and vibrations, the crankshaft is subjected to very high and complicated loading. Its design must, therefore, take all these problems into account. Overall load and maximum engine speed primarily determine the number of crankshaft bearings. Most diesel engine and high-speed spark-ignition engine crankshafts have a bearing after each throw due to the high working pressures. Four cylinder engines generally use a five bearing crankshaft and six cylinder engines use a seven bearing crankshaft. The crankshafts of some smaller engines that are not so highly loaded have a bearing only after every other crankshaft throw to reduce costs	Crankshaft materials are similar to those used for the connecting rods as described above
Crankcase or cylinder block	The cast crankcase is standard equipment on all motor vehicle engines. Its main purpose is to support the crankshaft, con-rod and piston assemblies. In many SI engines, the pistons run in cylinder liners cast as part of the main block. The crankcase often extends underneath the centre of the crankshaft to increase its strength. Most of the external engine components are also mounted on the crankcase	Crankcases for these engines were traditionally made from cast iron but are increasingly being made of aluminium in order to reduce engine weight. In diesel engines, separate wet or dry liners made of special wear-free materials are often used
Valves, guides and seats	The valves, guides and seats are designed to make a gas-tight seal against combustion pressures and temperatures. The valve guides of high-performance engines must have high thermal conductivity and good anti-friction properties. To improve the sealing of valves, rotating systems can be used. These mechanisms impart a rotary motion to the valves. Valve guides can be pressed into the cylinder head. Some modern engines have valve guides that are an integral part of the cylinder head. Both inlet and exhaust valves have valve-stem oil seals on their cold ends in order to reduce oil consumption	Valves are made of heat- and scale-resistant materials. Often the valve seat area is hard-faced. Thermal conductivity of the exhaust valve can be improved by filling the hollow stem with sodium. Valve seats are made out of cast or sintered materials and are fitted into the cylinder head by shrinking. This generally reduces valve seat wear. Hardened seats are required when using lead-free petrol
Cylinder head	The cylinder head seals the crankcase and the cylinders at the top. It houses the valves, spark plugs and/or injectors. In conjunction with the piston, it forms the combustion chamber shape. In the majority of car engines, the valve timing gear is also mounted in the cylinder head. There are two basic designs: 1. Counter-flow, where the inlet and exhaust valves are located on the same side of the cylinder head.	Cylinder heads of nearly all passenger car SI and diesel engines are now made of aluminium for reasons of weight and thermal conductivity

Name	Functional requirements	Materials
	2. Cross-flow, where the inlet and exhaust valves are located on opposite sides of the cylinder head	
Camshaft	The purpose of the cam is to open the valves as far as possible very quickly. This process should also be as smoothly as possible. The closing force for the valves is applied by the springs. The springs are also responsible for maintaining contact between the cam and the valves. Remember that the camshaft rotates at half engine speed	Camshafts are broadly made of either forged or cast steel. The individual cams are hardened. Many sophisticated techniques are used for this process

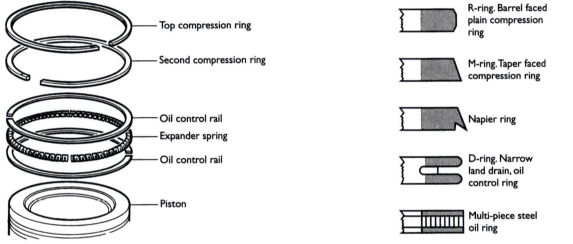

Figure 4.4 A common layout for piston rings

Figure 4.5 Piston ring shapes (in section)

- Top compression ring
- Second compression ring
- Oil control rail
- Expander spring
- Oil control rail
- Piston

- R-ring. Barrel faced plain compression ring
- M-ring. Taper faced compression ring
- Napier ring
- D-ring. Narrow land drain, oil control ring
- Multi-piece steel oil ring

LEARNING TASKS

➡ Look back at the key words. Explain each one to a friend, and/or write out a short description to keep as evidence.

➡ Make a simple sketch to show two different piston ring arrangements.

➡ Write a short explanation about why the materials used for engine components is so important. Give examples to show what you mean.

---

3 Engine design

Valve timing   Figure 4.6 shows a valve timing diagram showing an example of when the valves open and close. At first view, this is quite straightforward but, on closer examination, further points may need an explanation.

Multi-valve engines   The type of valve timing-gear assembly is mainly determined by the combustion-chamber shape. Most valve timing gear assemblies are of the overhead camshaft design, mounted in the cylinder head (Figure 4.7).

In earlier CI and SI engines subject to lower loads, the valves are parallel to the cylinder axis, and are usually actuated by bucket tappets or rocker arms. On modern spark-ignition engine designs, where performance and economy are the main considerations, valves are often tilted toward one another. This method, with a given cylinder diameter, allows larger valve diameters and better routing of inlet and exhaust passages.

KEY WORDS

- Balance
- Torsional vibration
- Valve overlap
- Valve lead
- Valve lag
- Compressor
- Turbine
- Wastegate
- Multi-valve engines

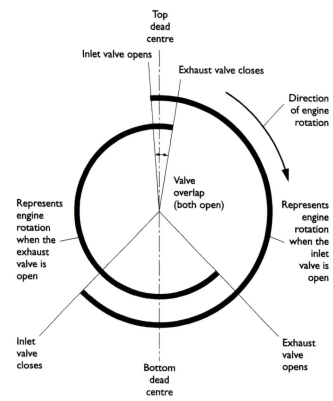

Figure 4.6 Valve timing diagram

Valve	Lead	Lag	Overlap
Exhaust	This maximises the expulsion of the exhaust gas. When the valve opens, the gas will still be at a pressure of 3 or 4 bars, so it will start to exit even though the piston is moving down the bore. The piston now only has to sweep the remaining gas out on its exhaust stroke	This causes an overlap with the inlet valve. A small amount of exhaust gas left in the new charge is known as 'internal exhaust gas recirculation' or EGR. This cools the burning a little and can reduce NOx emissions but can increase CO and HC emissions	The exhaust valve is left open after the start of the induction stroke because the outgoing gases leave behind a partial vacuum. This overlap with the inlet valve helps to 'draw' in the new mixture, even though the inlet valve is only partially open at this stage
Inlet	This causes an overlap with the exhaust valve. It is to allow the maximum possible time for the inlet charge to enter the cylinder	Inlet valve lag utilises the momentum (desire to keep moving) of the incoming charge. This is sometimes also referred to as inertia. The extra open time allows more charge to enter the cylinder. This improves volumetric efficiency	To get as much new mixture into the cylinder as possible, the inlet valve is opened before the exhaust valve has closed, so as to cause an overlap. However, if the valve is opened too early then exhaust gas could be forced out of it into the induction system

High-performance engines now often use four valves per cylinder with overhead bucket-tappet valve assemblies. A greater number of valves in a cylinder makes production a little more difficult, but the increased size of opening improves volumetric efficiency. The spark plug can also be placed in a central position. This has advantages for combustion development. A further advantage is that two cams are often used, allowing the timing of the cams to be changed with respect to each other.

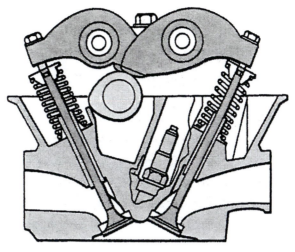

Figure 4.7 Four valves per cylinder operated by single camshaft and adjustable rockers

Turbo chargers   A turbocharger is an air pump that takes in air at ambient temperature, compresses it to a higher pressure, and then delivers this increased pressure to the engine intake system (Figure 4.8). Engine exhaust gases drive a turbine, which in turn drives a compressor wheel. The increased density of air into the engine permits better filling of the engine cylinder, thus increasing engine power output, whilst maintaining the correct air-fuel ratio for efficient combustion (14.7:1). Turbochargers are precision-built units. The shaft assembly usually revolves between 1000 and 130 000 rev/min. A device

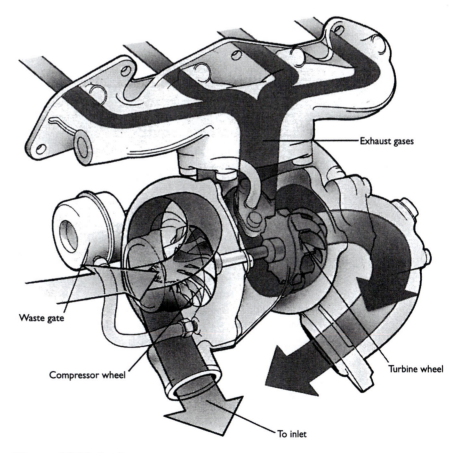

Figure 4.8 Turbocharger

called a wastegate is used to control the maximum boost pressure. The advantages of fitting a turbocharger are as follows:

- engine power output can be increased, by up to 30%, without increasing the engine speed or displacement
- utilising exhaust energy improves performance and increases fuel efficiency
- reduced emissions
- lower exhaust system noise
- increased engine torque giving better vehicle acceleration and pulling power.

A normal petrol engine only uses about 30% of the energy contained in the fuel; the remaining 70% of this energy is lost as follows:

- 37% heat energy to the outgoing exhaust gases
- 17% heat energy to the engine's coolant system
- 9% heat energy to the surrounding air
- 7% heat energy to overcome friction, pumping and component movement.

A turbocharger uses some of the energy contained in the exhaust gas to drive a turbine, which in turn drives a compressor wheel. The turbocharger extracts roughly up to a third of the wasted energy passing out from the engine's exhaust. The one problem, however, is that this increases the manifold back pressure. This makes it more difficult for spent gases to be expelled from the cylinders. Turbocharged engines produce better cylinder volumetric efficiencies compared with the normally aspirated engines. This causes higher peak cylinder pressures which increase the stress on engine components and, in petrol engines, could cause detonation. To overcome this problem, it is usual to reduce the engine's compression ratio. A normally aspirated engine with a compression ratio of 10:1 is often reduced to 9:1 for a low boost, or even to 8:1 if high boost pressure is used.

Under light load and low engine speed conditions, the energy in the exhaust gases will be small. At this level, it will not drive the turbine at very high speeds. This creates very little boost pressure and does not make any noticeable improvement to the engine's torque and power output. Therefore, because of its reduced compression ratio, a turbocharged engine can have lower torque and power outputs than a normally aspirated engine. A further problem with turbochargers is the delay or lag. When the engine is suddenly accelerated, there will be a small time delay before the extra energy discharged into the exhaust can speed up the turbine.

Power outputs of engines are tested and rated at sea level where the air is most dense. At increased altitude, the air becomes less dense. Less air will therefore be drawn into the cylinders, which leads to a reduction in engine power. A normally aspirated engine can have its power output reduced by approximately 14% when 1000 m above sea level. Turbocharged engines will still lose some power at high altitudes, but the loss will only be about 8%. The reason for this is that the turbine speed increases in line with the pressure difference between the exhaust gas entering and leaving the turbine. The exit pressure is the same as surrounding air pressure. The result is that at higher altitude the turbine will run faster, increasing the compressor speed.

A turbocharger consists of three main components.

- The compressor wheel is an aluminium alloy casting. This takes the form of a disc mounted on a hub with radial blades projecting from one side.

- The turbine is usually made from a high temperature heat resistant, nickel-based alloy such as 'Inconel'. This is because the exhaust gas temperature at the inlet to the turbine ranges between 600°C to 900°C.

- The wastegate is a device to limit maximum boost pressure. This is achieved by diverting some of the exhaust gas past the turbine. One of the most common methods is to use the boost pressure to open a valve. A pipe from the inlet allows the pressure to act on a diaphragm, which in turn opens the bypass valve or wastegate at a pre-set level. Some engines control the wastegate electronically as well as by boost pressure.

The turbine and compressor wheels are shown as Figure 4.9.

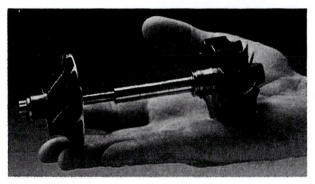

Figure 4.9 The heart of turbocharging technology. The exhaust gases power the turbine wheel on the right, so that the impeller on the left presses more air into the engine. This increases the amount of power and torque the engine can provide from the same cylinder volume

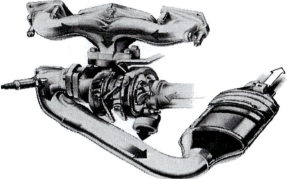

Figure 4.10 The three-way catalytic converter is fitted close to the engine so it quickly reaches its working temperature

Engine balancing   Bending vibrations are significant on engines with a small number of cylinders, because the crankshaft and the large flywheel cause low frequency oscillations. This applies bending and torsional vibrations to the crankshaft. However, even with increasing numbers of cylinders, the torsional vibrations of the crankshaft, connecting rod and piston, can be dangerous. The natural frequencies and vibrational impacts can be determined so that suitable crankshaft dampers can be used to reduce the effects. The effects of an unbalanced engine fall mainly into just two categories:

- unpleasant vibrations
- severe mechanical damage.

The second of these two clearly requires immediate action. The torsional vibrations of the crankshaft must be reduced to uncritical values through the use of vibration dampers. Bonded rubber vibration dampers or viscous vibration dampers are the most common. These usually form part of the front crankshaft pulley. In some cases, counterbalance shafts are used. These shafts are gear driven from the crank and are carefully weighted to produce vibrations, which are equal and opposite to those caused by the crank and pistons, thus cancelling their effect.

LEARNING TASKS

➡ Look back at the key words. Explain each one to a friend, and/or write out a short description to keep as evidence.

➡ Make three simple sketches using valve-timing data to compare different engines.

➡ Examine a real system and note how components are balanced.

➡ Write a short explanation about why turbochargers need some form of wastegate.

4   The 'K' series engine case study

Introduction   The 'K' series engine (Fig 4.11) has gained a reputation for quality and reliability. It was launched in the Rover 200 range and then introduced into the 400 series and across the new Metro range.

Figure 4.11 Rover 'K' series engine

The original design objectives were simple. The engine was to have excellent levels of performance and economy, but still operate within existing and proposed pollution laws. 'K' series achieved all of these targets.

During the development period, it became clear that tougher pollution legislation was destined to be introduced in Europe. The 'K' series engine easily met the new requirements and was available with an exhaust gas catalyst at launch.

The more stringent emissions legislation in Europe meant the new engine needed an overhead camshaft with the option of multi-valve combustion chambers. Different combustion chamber designs were manufactured and flow tested on special rigs. The best of these designs were then made in alloy and tested on a development engine. The 'K' series engine, as a result of this development, is able to maintain stable combustion, at air/fuel ratios greater than 21:1. This is well into the lean-burn range.

'K' series general construction

All major castings on the 'K' series engine are made from an aluminium alloy. This material is much lighter than cast iron. Its one major disadvantage, however, is a lack of tensile strength. Previously, engineers solved this problem by increasing the wall thickness of the castings. This, however, reduced the weight advantage. Rover engineers found an ingenious solution in a revolutionary new 'sandwich' build technique.

KEY WORDS

■ Wet liners
■ Hydraulic tappets
■ Through bolts
■ Combustion chamber profiles

There are three main components of the engine:

■ cylinder head
■ cylinder block
■ main bearing ladder.

'K' series range   The 'K' series range consists of three engines:

Engine designation	1.4 K16	1.4 K8	1.1 K8
Camshaft	Twin	Single	Single
Valves per cylinder	4	2	2
Capacity	1397 cc	1397 cc	1120 cc
Bore	75 mm	75 mm	75 mm
Stroke	79 mm	79 mm	63.25 mm
Compression ratio	9.5:1	9.75:1	9.75:1
Fuel system	MEMS SPI	KIF 44 carburettor	KIF 38 carburettor
Ignition system	MEMS programmed	Constant energy	Constant energy
Maximum power	95 PS @ 6250rpm	76 PS @ 5700rpm	60 PS @ 6000rpm
Maximum torque	91 lb.ft @ 4000rpm	86 lb.ft @ 3500rpm	66 lb.ft @ 3500rpm

These components make up three layers of the sandwich. They are held together in compression by long bolts which extend right through the assembly and screw into a small ladder casting at the base. Thus, a single set of fixing bolts is used for the cylinder head and main bearing caps. Apart from the benefit of a reduction in fixings, this technique helps to keep the weight of the 'K' series down.

Aluminium alloy has relatively low tensile strength, but it is very strong when subjected to compression. This means that by using high tensile steel through bolts to hold the engine together, it is possible to design casting sections to withstand only compression forces. The main castings of the 'K' series engine are therefore much thinner and lighter than many of their counterparts.

Wet liners and block construction   An interesting feature of the 'K' series engine is that it uses 'wet' cylinder liners (Figure 4.12). This type of liner is termed 'wet' because it is surrounded by coolant. In most petrol engines, coolant circulates around passages in the block and does not come into direct contact with the liners. This means heat dissipation from the cylinder walls can be a relatively slow process. The use of wet liners ensures an even cooling effect around the cylinder bores that, in turn, prevents distortion.

Metering holes in the cylinder head gasket controls coolant flow around the liners. The liners are a slide fit in the block, and are held in place by a small lip at the top, which is clamped between the cylinder head and block. As the engine warms and cools, the 'top hung' liners are free to expand and contract at their own rate. As a result, the liners are not affected by the different expansion rates of the surrounding castings and remain in perfect shape. This improves engine life and gives excellent oil consumption. The liners have two rubber 'O' ring seals at the base to prevent coolant leakage into the crankcase.

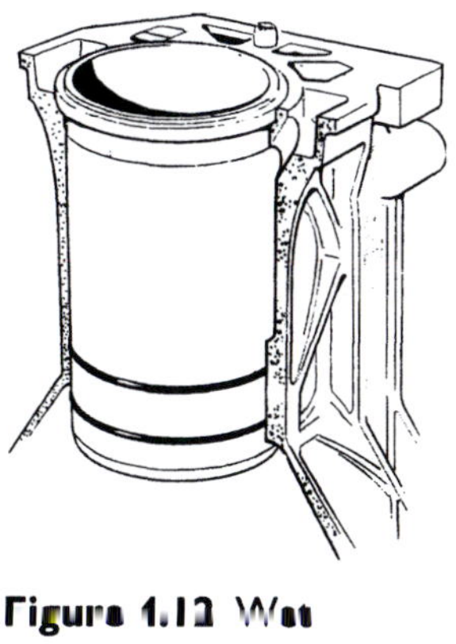

The 'K' series engine uses a five main bearing, cast iron crankshaft with eight balance weights. End float is controlled by thrust washers positioned either side of the centre main bearing in the cylinder block. A bearing ladder instead of the more usual separate bearing caps secures the crankshaft. The bearing ladder is common to all 'K' series engines. The 'rungs' of the ladder

Figure 4.12 Wet cylinder liners

form the lower halves of each main bearing. This technique improves structural rigidity, which leads to lower vibration levels. The block and bearing ladder are clamped together and line bored, making them a matched pair. The bearings themselves are conventional white metal shells.

Number 2 and 4 main bearing shells are grooved, which oil from the main gallery is circulated through and then channelled through drillings in the crankshaft to the big-end journals. All the bearing shells are colour coded to indicate their thickness grade.

The crankshaft tightens when the through bolts are removed because there is virtually no clamping load on the main bearings (Figure 4.13). This causes a small distortion of the bearing bores, which in turn makes the crankshaft tight. This is normal, but do not rotate the crankshaft any more than is necessary until the through bolts are fitted.

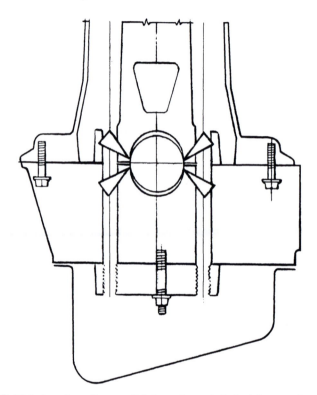

Figure 4.13 Main bearings distort slightly and crankshaft tightens when the through bolts are removed

Pistons   There are three different types of piston used in the 'K' series engine range, one for each engine derivative. Each piston uses two compression rings and one oil control ring. The top compression ring is plain and a stepped ring is used in the second groove. The oil control rings differ between K8 and K16 engines. The K8 uses conventional three-piece rings and the K16 piston uses a two-piece oil control ring.

The 'K' series engine uses semi-floating gudgeon pins that are a slide fit in the piston and an interference fit into the connecting rod. All 'K' series pistons have a gudgeon pin offset towards the 'thrust' side of the engine (left hand side when viewed from the crankshaft pulley end). This is done to counteract thrust forces (Figure 4.15).

The oil gallery is an important component in the 'K' series engine, because:

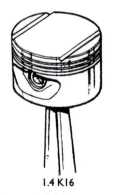

1.4 K16

Figure 4.14 Piston used in the 'K' series 1.4 K16 engine

- it distributes oil to the main bearings
- it acts as a 'nut plate' for the through bolts
- it provides further strengthening for the main bearing ladder.

Cylinder heads Two different cylinder heads are used across the 'K' series range: a 16 valve, twin overhead camshaft head is used on the K16, whilst both K8 engines have a common single overhead camshaft 8 valve head. The valves are operated through hydraulic tappets. Both cylinder heads are of the cross-flow design where fuel/air mixture enters on one side of the cylinder head and leaves as exhaust gases on the other.

The cylinder heads, and combustion chamber profiles, take much of the credit for 'K' series' performance, economy and low exhaust emissions. The effects of combustion chamber profiles and valve configurations on engine performance and economy are significant. Research showed that if a barrel swirling motion was induced in the inlet charge, a faster combustion rate could be attained. This fast burn rate is an essential requirement in lean burn engines.

'Barrel swirl' is the term used to describe turbulence inside the cylinder, which swirls across the axis of the bore rather than around it. Turbulence around the axis of the bore is called 'axial swirl'.

The shape of the inlet port and the opposing wall of the combustion chamber both force the intake charge downwards, which starts the barrel swirl. The barrel swirl continues as the piston descends on the intake stroke. On the compression stroke, the barrel swirl is forced into a smaller volume and increases in velocity. Eventually, the space becomes too confined for the swirl to take place and it breaks down into hundreds of smaller 'micro swirls' (Figure 4.16). These micro swirls thoroughly mix the intake mixture, which results in fast, even combustion. The 'K' series cylinder head, in 8 and 16 valve forms, is fully serviceable.

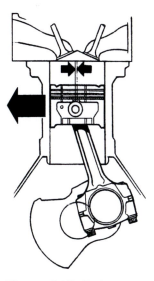

Figure 4.15 Gudgeon pin is offset to counteract thrust forces

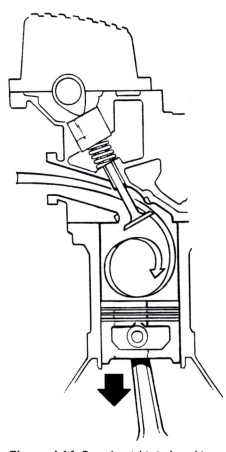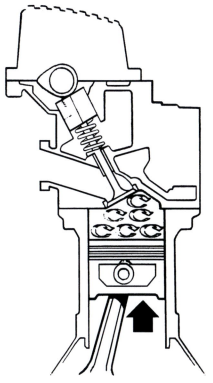

Figure 4.16 Barrel swirl is induced in the charge of intake mixture

The swirl velocity increases and finally breaks up into smaller 'micro' swirls

The 'K' series engine uses a stainless steel backbone gasket with moulded silicone beads for fluid sealing and flame rings for combustion gases. This high-tech gasket was developed on the Formula 1 race circuit and gives far better sealing qualities than conventional laminated designs. Compression limiters are fitted at each end of the gasket to prevent over compression of its sealant beads.

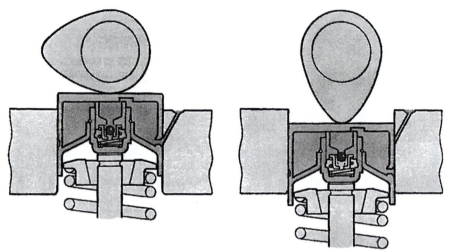

Ball valve opens and low pressure oil fills lower chamber

Ball valve closes as valve opens

Figure 4.17 Hydraulic tappets

Hydraulic tappets

Hydraulic tappets are used to automatically adjust valve clearances (Figure 4.17). The actual operation of the valves, however, is still mechanical. The main body of the tappet is positioned between the valve stem and camshaft as with normal tappets. Inside the tappet, there is a small spring-loaded piston with a ball valve. Oil is fed to the tappet at low pressure from the engine oil pump. When the camshaft starts to open a valve, the tappet ball valve snaps shut and seals the small high-pressure chamber inside the piston. This creates a powerful hydraulic lock, which remains for the full duration of valve opening. When a gap develops between the tappet and camshaft, a spring makes the tappet piston extend from its body until the gap is closed. This causes an increase in volume in the high-pressure chamber, which is filled by additional oil drawn through the ball valve. The high-pressure chamber remains fully primed for long periods and only occasionally needs to be recharged.

Through bolts

The through bolts used in the 'K' series engine are a vital component. Figure 4.18 shows the difference between long through bolts and the conventional clamping method. It is therefore *essential* to check their condition if they to be re-used during an engine rebuild. In many cases, it is best to renew them if in any doubt. As a result of their length and double role, the 'K' series through bolts have an unusual tightening sequence. It is most important to follow the correct sequence or the head face may become distorted.

- Tighten the through bolts in normal rotation to 20 Nm.
- Mark cylinder head adjacent to line on bolt flanges.
- Tighten each bolt in sequence approximately 180°.
- In sequence, tighten bolts a further 180° to align marks.

The cylinder head can be removed without disturbing the bottom end because, when the through bolts are removed, the smaller auxiliary fixings hold the bearing ladder and oil gallery in place. However, without the correct clamping load on the bearing ladder, the crankshaft will tighten.

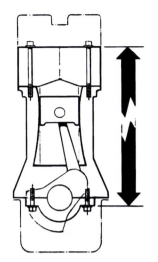

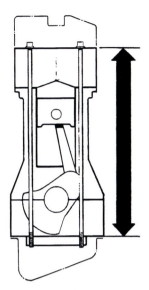

Figure 4.18 Conventional clamping arrangement can lead to distortion of the cylinder bores

'K' Engines – Long through bolts hold the main castings together and bore distortion is avoided

Engine breathing

Figure 4.19 The 'K' series engine uses its through bolt apertures for crankcase breathing and oil drain

The 'K' series engine breather system is almost completely internal (Figure 4.19). Passageways formed by the through bolt construction are used instead of external pipes and separators. Each through bolt has its own passageway that extends all the way through the engine. Because the engine is inclined forwards by about 15°, oil from the head will drain back into the crankcase through the front through bolt holes. The rear through bolt holes are therefore able to carry crankcase fumes. Separating oil drain and crankcase fumes in this way guards against oil being pulled over into the intake system.

On K16 engines, two small pipes from the camshaft cover carry crankcase fumes to the throttle body, either side of the throttle disc. The camshaft cover gasket doubles as a baffle plate to prevent oil splash from the camshafts entering the breather system. The K8 camshaft cover has an integral baffle chamber and crankcase fumes are fed through a single pipe into the carburettor.

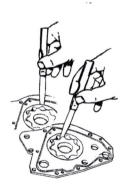

Figure 4.20 Remove the oil pump rear cover to check rotor clearances

All of the 'K' series engines use a multi-lobe eccentric rotor type oil pump (Figure 4.20). Flats on the crankshaft drive the oil pump. The pump itself is not serviceable, but it is possible to separate the two halves of the pump to check for wear. It has the following specifications:

- oil pressure at idle should be at least 15 lbf/in^2
- oil pressure at 3000 rev/min should be at least 55 lbf/in^2
- if oil pressure is very high at 3000 rev/min, suspect the oil pressure relief valve
- outer rotor to housing gap – 0.28 to 0.36 mm
- inner rotor tip gap – 0.05 to 0.13 mm
- rotor end-float – 0.02 to 0.06 mm.

Camshaft drive

All 'K' series derivatives use glass fibre reinforced, round toothed, rubber timing belts (Figure 4.21). The purpose is to transmit drive from the crankshaft to valve gear. Advancements in drive belt technology have enabled the service life to be pushed as far as 96 000 miles with no tension

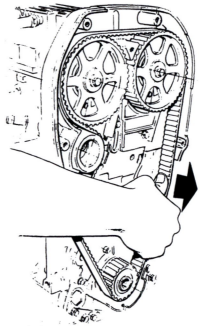

Figure 4.21 Checking cambelt tension

checks in between. The important tasks to perform on this, or any other, belt during normal servicing are listed below. You should check for:

- splits
- excessive wear
- contamination.

Remember to always mark the direction of rotation on the belt prior to removal if you plan to refit it. Industry practice, however, is to replace the belt with a new one.

Conclusion The 'K' series engine has now been in use for a number of years and has proved to be a reliable engine. Significant developments were made during its design and construction. Many of the points highlighted in this case study are also relevant to some other engines but we should give credit where it's due.

LEARNING TASKS

➡ Look back at the key words. Explain each one to a friend, and/or write out a short description to keep as evidence.

➡ Make a simple sketch to show what is meant by 'barrel swirl'.

➡ Examine a real system, if available, and note the construction techniques and materials used.

➡ Write a short explanation about what you consider to be the advantages and disadvantages of the 'K' series engine.

---

5 The MDi diesel engine – case study

Introduction Figure 4.22 shows a current diesel engine from the Rover range. However, because other relevant issues relating to diesel fuel systems and general engine technology are covered elsewhere, this case study will examine an older engine. This will help us to cover other important issues.

Figure 4.22 One of the latest Rover diesel engines

The MDi diesel is a well-known, tried and tested direct injection engine. The main internal components of the MDi engine are shown as Figure 4.23. This case study will highlight some interesting and fundamental issues relating to light vehicle diesel engines, their build and operation. A section is included relating to the smoke produced by a diesel engine. This will be relevant to all engines, not just the one described here. The engine is used in a range of cars and light vans. A turbocharged and normally aspirated version are also available. I have not attempted to describe every aspect of the engine – but the important issues are covered.

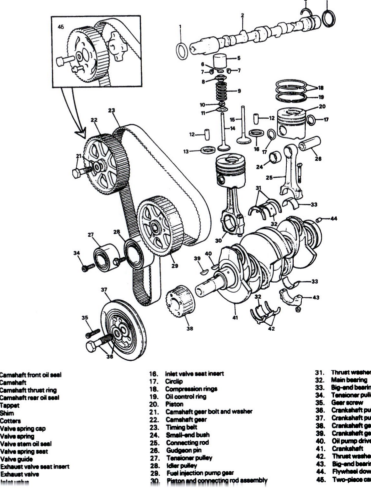

1.	Camshaft front oil seal	16.	Inlet valve seat insert	31.	Thrust washer—top half
2.	Camshaft	17.	Circlip	32.	Main bearing
3.	Camshaft thrust ring	18.	Compression rings	33.	Big-end bearing
4.	Camshaft rear oil seal	19.	Oil control ring	34.	Tensioner pulley screw
5.	Tappet	20.	Piston	35.	Gear screw
6.	Shim	21.	Camshaft gear bolt and washer	36.	Crankshaft pulley bolt and washer
7.	Cotters	22.	Camshaft gear	37.	Crankshaft pulley
8.	Valve spring cap	23.	Timing belt	38.	Crankshaft gear
9.	Valve spring	24.	Small-end bush	39.	Crankshaft gear drive key
10.	Valve stem oil seal	25.	Connecting rod	40.	Oil pump drive key
11.	Valve spring seat	26.	Gudgeon pin	41.	Crankshaft
12.	Valve guide	27.	Tensioner pulley	42.	Thrust washer—lower half
13.	Exhaust valve seat insert	28.	Idler pulley	43.	Big-end bearing cap
14.	Exhaust valve	29.	Fuel injection pump gear	44.	Flywheel dowel
15.	Inlet valve	30.	Piston and connecting rod assembly	45.	Two-piece camshaft gear

Figure 4.23 Internal engine components

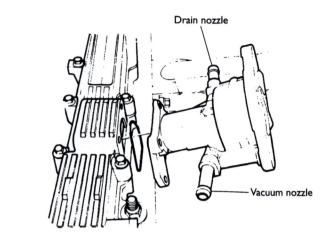

Drain nozzle

Vacuum nozzle

Figure 4.24 Braking system exhauster

Engine ventilation

The MDi engine is designed to run with a slight positive pressure in the crankcase. This helps to remove fumes more quickly. The breather unit is attached to the oil filler tube and will control the pressure to the correct level. Any excess is vented to the air intake system through a pipe. On overrun, minimum fuel is injected but air is still taken into the engine. The vent valve is 'sucked' to the closed position to prevent oil mist from the crankcase being taken into the engine. The diesel engine uses oil as a fuel and if enough oil mist was taken into the engine the speed could increase even with the throttle closed.

Braking system servo exhauster

The exhauster (Figure 4.24) is necessary, as there is no throttle valve or venturi to create a sufficient depression necessary for the servo operation. The throttle on this engine in common with most other diesels only controls the fuel via the rotary injection pump. The exhauster is driven from the camshaft and uses a piston to produce the pumping action. The exhauster used on this system should produce a minimum vacuum at idle of 75 kPa (22 in. or 560 mm of Hg). If you prefer absolute pressure as I do, 75 kPa vacuum is about 25 kPa absolute!

Camshaft drive

Two types of camshaft gear and gear location to the camshaft are used. The early type is a one-piece pulley that is not keyed to the camshaft. Drive is transmitted by the engagement of the gear onto the end of the camshaft where the centre bolt torque ensures the friction drive rotates the camshaft. A two piece pulley is fitted to all later engines, both turbocharged and naturally aspirated. The pulley hub is located to the camshaft by a peg and retained by a normal centre bolt. Four screws are used to secure the gear to the hub via four slots.

Timing belt

Any timing belt must be handled with care. Do not use a belt that is contaminated with oil or that has been bent at a sharp angle. This could fracture the reinforcing fibres and cause failure. Timing belts should be stored on edge and in a circular shape – not hanging up on pegs. During normal operation, a timing belt develops a wear pattern. If it is to be reused the direction of rotation should be marked on the belt's outer face with chalk. If the belt teeth show signs of wear, it must be renewed. The outer face of the belt should also be inspected as most of the belt's strengthening fibres are close to the outer face. The MDi engine is fitted with a 30 mm wide belt for long life, low noise and accurate timing. It also has arrow marks on the outer face to ensure it is fitted correctly. Service life is 96 000 miles with a visual inspection at 48 000 miles.

If any sign of damage is noticed, or if in any doubt, then renew the belt.

The tension of a belt should be checked using a tension gauge. Care must be taken to use the belt tension gauge in the correct way. The following list is a general guide.

- Extend the hook to the extreme position.

- Slip the belt between nose-piece and hook; be sure the nose-piece is centred over centre of belt.

- Release the handle with a rapid movement. A slow release will result in a high reading because friction of the belt on the gauge will prevent the hook from retracting fully.

- Read units off the dial.

Cylinder block and cylinder head

The cylinder block is made from high-grade cast iron. It was designed using computer-modelling techniques. This resulted in the maximum weight reduction and addition of struts designed for enhanced rigidity and low noise characteristics. The cylinder block can be rebored oversize by 0.50 mm if required. The main oil gallery is drilled to allow the fitting of piston cooling oil jets on turbocharged engines. On naturally aspirated engines, blanking plugs are fitted instead of the oil jets.

The aluminium cylinder head can be overhauled like most other types. It can either be removed by first removing the cam cover and camshaft or as a complete assembly. An important point to note is that you should protect the aluminium head face against damage by laying the head on its side or on wooden blocks. Also, take care of protruding valve heads. If the head is removed, a gasket-releasing agent should be used for cleaning, rather than scraping the gasket material away. Bow can be checked with a straight edge and feeler gauges in the usual way. The maximum permissible bow of the MDi cylinder head is 0.10 mm. It can be machined by a maximum of 0.2 mm.

Piston and connecting rods

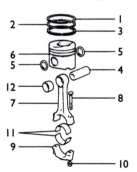

1. Top compression ring
2. Second compression ring
3. Oil control ring and spring
4. Gudgeon pin
5. Circlips
6. Piston
7. Connecting rod
8. Connecting rod bolt
9. Connecting rod cap
10. Self-locking nut
11. Big-end bearings
12. Small-end bush

Figure 4.25 Piston and connecting rod

The turbocharged engine is fitted with piston cooling jets fed from the main oil gallery. As well as the banjo bolt securing each jet, a dowel ensures the correct jet angle and clearance to the cut-outs in the piston skirt is maintained.

The three-ring aluminium 'bowl-in' pistons (Figure 4.25) have steel inserts for controlled expansion. These help to reduce noise and oil consumption. An armoured top ring groove is used to give longer life to the piston. The top compression ring will take most of the high pressures in the diesel engine cylinder.

The piston is secured to the con-rod with a fully floating gudgeon pin retained by circlips. Con-rods must be fitted to the pistons correctly as the gudgeon pin is offset to the thrust side of the engine.

The top ring has an internal chamfer, which must be uppermost. The second ring is a tapered type. Both compression rings are marked 'TOP'. A two-piece oil control ring is used; this can be fitted either way up. New compression rings are colour coded on the outside face. As a guide to correct fitment, the paint mark must be on the left of the ring gap when the piston is held the correct way up. The top ring is red and the second ring green. Piston ring gaps (new cylinder bore) should be as follows:

- compression rings – 0.28 to 0.56 mm

- oil control ring – 0.23 to 0.56 mm.

In common with most engines using three piston rings, the ring gaps should be spaced at 120° intervals. When fitted, the piston protrusion must be within 0.46 to 0.65 mm.

Cylinder compression tests

Cylinder compression test results should not be used in isolation but to confirm other diagnostic evidence. Diesel engine cylinder compression pressures should generally be in the range of 20–35 bar. Readings can be affected by:

- the state of the battery charge
- efficiency of the starter motor
- ambient temperature
- the type of gauge used.

For these reasons, it is difficult to quote precise figures, and you should be looking for a cylinder that is different from the remainder. Any cylinder that differs by more than 3.5 bar from the others should be regarded as suspect. The following list is a good guide to a checking procedure.

- Ensure valve clearances are within tolerance.
- Run engine until it reaches normal operating temperature.
- Disconnect wire from stop/start solenoid and crank the engine a few turns to empty the high-pressure fuel chamber in the fuel injection pump.
- Remove all injectors, seat washers and location rings.
- Using a seat washer from an injector, fit and secure gauge to one cylinder.
- Crank engine until gauge pressure stabilises and record pressure.
- Repeat for other cylinders.

Crankshaft

The crankshaft has five main bearings. It has deep, cold rolled radii on all journals for increased strength and rigidity. Grooved shell bearings are fitted to the crankcase and plain bearings to the main bearing caps. The centre main bearing is wider than the other four.

Four thrust washers control crankshaft endfloat. These are fitted each side of the centre main bearing. The two thrust washers in the bearing cap are located with tags to prevent the thrust washers revolving around the journal. Crankshaft end-float should be in the range 0.3 to 0.26 mm. All crankshaft journals can be reground to 0.3 mm undersize if required.

Lubrication system

The engine uses a pressure fed, wet sump full-flow system. It has a disposable canister oil filter containing a bypass valve. A pressure relief valve (Figure 4.26), fitted in the oil pump housing, controls oil pressure. The

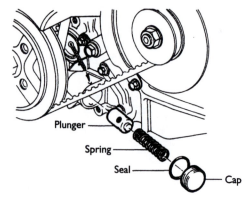

Figure 4.26 Oil pressure relief valve

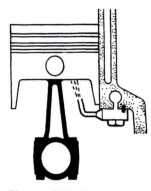

Figure 4.27 Piston cooling jets

high capacity oil pump is a multi-lobe eccentric rotor type. It is driven directly off the crankshaft nose using a woodruff key.

Oil is drawn from the sump, through a gauze strainer and a passage in the cylinder block to the oil pump. It then passes the pressure relief valve to the full-flow oil filter. The oil filter contains a bypass that opens if the filter element becomes choked. This will mean that unfiltered oil is passing around the engine but even this is better than no oil at all. From the filter, oil passes to the main gallery. From the gallery, drillings feed oil to the crankshaft main bearings, the cylinder head and oil cooling jets on a turbocharged engine. Drillings in the crankshaft feed oil to the big-end journals. In the cylinder head, oil-ways feed oil to all the camshaft journals via the housing.

Figure 4.27 shows how oil jets are used to provide extra piston cooling on the turbocharged engine.

Fuel system

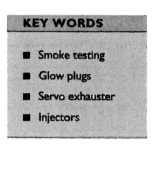

KEY WORDS

- Smoke testing
- Glow plugs
- Servo exhauster
- Injectors

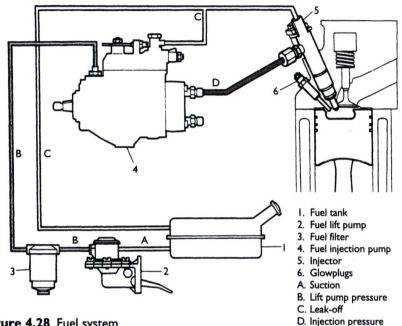

1. Fuel tank
2. Fuel lift pump
3. Fuel filter
4. Fuel injection pump
5. Injector
6. Glowplugs
A. Suction
B. Lift pump pressure
C. Leak-off
D. Injection pressure

Figure 4.28 Fuel system

Figure 4.28 shows the general layout of the MDi engine fuel system. The component parts of this system will be examined later, but first let us look at the general background on diesel fuel.

Fuel oils for use in diesel engines are generally known as Derv, Derv fuel, distillate diesel fuel or automotive gas oil. Because they need to be suitable for a wide range of operating conditions, diesel fuels are complex compounds. Summer and winter grades with different low temperature properties are produced. A British Standard specifies the temperature above which the fuel flow rate must meet certain criteria. This defined temperature is known as the 'Cold filter plugging point' (CFPP). Below this temperature 'fuel waxing' can occur. This is the formation of wax crystals, which will restrict fuel flow by partially blocking the supply system.

Symptoms similar to those of wax separation can result when fuel is contaminated by water, as ice crystals will form at temperature of 0°C and below.

Fuel additives

The fuel lubricates all of the injection pump and injector's internal components. Additives can reduce the lubricating properties of the fuel resulting in rapid wear or damage to both the engine and the fuel injection equipment. There are some fuel additives that are moderately beneficial in that they lower the CFPP, but these are only effective if correctly mixed into the fuel before wax separation occurs. Once wax separation has taken place, it can only be eliminated by physically removing the blockage or by warming the fuel to dissolve the crystals.

Kerosene is technically preferable to other additives but its use in road vehicles in the UK and certain other markets is illegal, as duty has not been paid. It would therefore be necessary to make special arrangements with Customs and Excise before using it as an emergency fuel; the exception in the UK being its use in vehicles not driven on public roads. If appropriate arrangements are made with Customs and Excise, up to 20% (1 part kerosene to 4 parts of diesel fuel) may be added during periods of extreme cold only.

The use of petrol as an additive is permitted, since duty has been paid but, in adding petrol to diesel fuel, the volatility of the mixture is increased and adequate safety precautions must be taken. Consequently, its use is not recommended.

Fuel tank

A conventional metal tank is used, but with an additional pipe for fuel to return from the injectors and injection pump leak-off pipes. The fuel filler pipe has a warning label to ensure only diesel fuel is put in the tank.

Fuel lift pump

A mechanical lift pump is fitted to the cam cover and is driven by an eccentric cam on the camshaft. It incorporates a hand-priming lever on the pump body. This is used when bleeding air from the fuel system. The internal gauze filter should be cleaned every 48 000 miles.

To check the operation of the pump, disconnect the pump outlet pipe and fit a 0–10 lbf/in^2 pressure gauge to the outlet union. Crank the engine for 10 seconds and note maximum pressure on the gauge. If pressure is less than 4.5 lbf/in^2, renew the pump. Also, check the time for the pressure to drop to half of the maximum figure obtained. If it is less than 30 seconds, renew the pump.

Fuel filter

The canister type fuel filter (Figure 4.29) contains a very fine element that prevents even the smallest particle of dirt entering the fuel injection pump. A drain cock is situated at the base of the filter, which is used for draining off water. This should be done at weekly intervals. If water is regularly present in the fuel filter, it could be an indication that there is a problem with the fuel storage tanks. If this is the case, water may be taken into the fuel tank every time you fill up.

To renew the filter, thoroughly clean its exterior, loosen the bleed screw, open the drain cock and drain the filter. Remove the filter and discard carefully. Check that the threaded adapter in the filter head is secure, then lubricate the seal of the new filter with clean fuel and fit using only your hands. With the bleed screw loosened, operate the priming lever on the fuel lift pump until bubble-free fuel emerges, and then tighten the bleed screw.

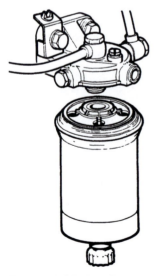

Figure 4.29 Fuel filter with a water drain tap

Fuel injection pump

Small, high-speed diesel engines require a light and compact fuel injection pump. The Bosch EPVE type fuel injection pump meets these criteria by combining a supply pump, mechanical governor, hydraulic timing device, electro-mechanical fuel shut-off mounting point and the high pressure pump chamber in a single compact body.

Other fitments include:

- a stop/start fuel cut-off solenoid for a key-operated stop/start of the engine
- a fast idle solenoid and push rod
- a cold start unit to advance the injection timing when starting from cold
- a torque control or smoke limiter, depending on whether the engine is naturally aspirated or turbocharged respectively.

The EPVE injection pump is a distributor type pump. This means it has only one pumping chamber regardless of the number of cylinders there are in the engine. As the pumping element moves longitudinally to generate high pressure, it is also rotated to deliver fuel to the hydraulic head in the firing order sequence.

The injection pump, according to engine speed, load and accelerator position, meters fuel. It then delivers this precise amount of fuel through the delivery valves, high-pressure pipes and injector into the combustion chamber. The MDi combustion chamber is a 'bowl-in' piston crown type combustion chamber.

The operation of the mechanical governor is proportional to engine speed and will control fuelling from 0 up to the engine rated speed of 4500 rev/min. Above its rated speed, the governor will gradually reduce fuelling, causing a gradual power loss. The governor is *not* the 'Brick Wall' type, which cuts off the fuel quickly.

Chapter 5 contains further details about fuel system operation.

Fuel injectors

If an injector (Figure 4.30) is suspected of being faulty, it can be tested before removing it from the engine. First, run the engine just above the low idle speed setting and loosen each injector union in turn and note if the engine speed changes. If it remains constant for any injector it means that the injector is faulty and must be renewed. This procedure is exactly the same principle as shorting out a spark plug on a petrol engine.

With the injector removed from the engine, a tester can be used to check the injector spray pattern. The MDi injector is a four-hole type. Each of the four sprays must be finely atomised and of good form, without any splits or heavy concentrations of fuel to one side.

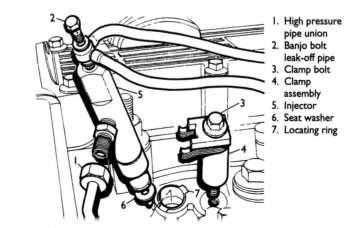

1. High pressure pipe union
2. Banjo bolt leak-off pipe
3. Clamp bolt
4. Clamp assembly
5. Injector
6. Seat washer
7. Locating ring

Figure 4.30 Injectors

The injector opens at 225 bar. Fuel at this pressure can easily penetrate the skin, so extreme care must be taken when working near high pressure fuel. To produce an extremely fast opening of the injectors, the high pressure system of the injection pump operates at 500–600 bar. Further wise precautions when working with diesel fuel is to wear thin protective gloves or apply barrier cream.

Glowplugs

Glowplugs are not often fitted to direct injection engines because fuel is injected into the hottest part of the cylinder – the cavity in the piston crown. However, fitting glowplugs to the direct injection MDi engine allows it to achieve near instant starting, even at temperatures well below freezing.

The MDi engine uses a very fast glowplug, which reaches a full operating temperature of 850°C within two seconds. Thereby lies the secret, together with direct fuel injection, of the excellent starting characteristics of this engine.

Due to the starting characteristics, it is not necessary for prolonged pre-heating before starting up. However, there is a short pre-heat period every time the ignition switch is turned from the 'off' to the 'on' position. The maximum pre-heat period is three seconds, and this is the only time the warning light on the instrument panel will glow. Both the pre-heat and post-heat periods are reduced when the engine is warm. The internal resistance, due to heating, within the glowplug determines the pre-heat period. An ambient sensor, situated within the controller, will sense the underbonnet temperature and control the post-heat accordingly. The four glowplugs (one for each cylinder) are wired in parallel. Figure 4.31 shows a typical glowplug.

Glowplug data:

- tip starts to glow after 5 seconds
- initial current draw is 20 amps
- current draw after 20 seconds is 12 amps.

Figure 4.31 Diesel engine glow plug

Diesel engine smoke

Diesel fuel is a hydrocarbon fuel. When it is burned in the cylinder, it will produce carbon dioxide and water. There are, however, many circumstances under which the fuel may not be completely burned and one of the results of this is smoke. Despite the fact that diesel engines are designed to run under all conditions with an excess of air, problems still occur. Very often, these smoke problems are easily avoided by proper maintenance and servicing of the engine and its fuel system. The emission of smoke is usually due to a shortage of air (oxygen). If insufficient air is available for complete combustion, then unburnt fuel will be exhausted as tiny particles of fuel (smoke).

The identification of the colour of diesel smoke and under what conditions it occurs can be helpful in diagnosing what originally caused it. Poor quality fuel reduces engine performance, increases smoke and reduces engine life. There are three colours of smoke: white, blue and black. All smoke diagnosis tests must be carried out with the engine at normal operating temperature.

White or grey smoke

White smoke is vaporised unburnt fuel and is caused by there being sufficient heat in the cylinder to vaporise it but not enough remaining heat to burn it. All diesel engines generate white smoke when starting from cold and it is not detrimental to the engine in any way – it is a diesel characteristic. Possible causes of white smoke are listed below.

- Faulty cold starting equipment. Cold engines suffer from a delay in the combustion process. A cold start unit is fitted to advance the injection timing to counteract this delay. This means that white smoke could be a cold start unit problem.

- Restrictions in the air supply. A partially blocked air cleaner will restrict the air supply; while this is an easy cause to rectify, it is often overlooked. Incidentally, a blocked air cleaner element at light load in the workshop becomes a black smoke problem when the engine is under load. In both cases, there will not be sufficient air entering the cylinder for the piston to compress and generate full heat for combustion.

- Cold running. Check the cooling system thermostat to see if the correct rated thermostat is fitted.

- Incorrect fuel injection pump timing. If fuel is injected late (retarded timing) it may be vaporised but not burned.

- Poor compressions. Poor compressions may lead to leakage during the compression stroke and, inevitably, less heat would be generated.

- Leaking cylinder head gasket. If coolant was leaking into the combustion area, the result would be less temperature in the cylinder, causing white smoke. Steam may also be generated if the leak is sufficient. All internal combustion engines have water as one of the by-products of burning fuel – you may have noticed this from your own car exhaust, especially on a cold morning.

Blue smoke

Blue smoke is almost certainly a lubricating oil burning problem. Possible causes of blue smoke are:

- incorrect grade of lubricating oil
- worn or damaged valve stem oil seals, valve guides or stems where lubricating oil is getting into the combustion chamber
- worn or sticking piston rings
- worn cylinder bores.

Black smoke

Black smoke is partly burned fuel.

Possible causes are:

- Restriction in air intake system. A blocked air cleaner element will not let enough air in to burn all the fuel.

- Incorrect valve clearances. Excessive valve clearances will cause the valves to not open fully and close sooner. This is another form of insufficient air supply.

- Poor compressions. Air required for combustion may leak from the cylinder.

- Defective or incorrect injectors. Check the injector to see if the spray is fully atomised and solid fuel is not being injected.

- Incorrect fuel injection pump timing. This is less likely because the timing would need to be advanced to the point where additional engine noise would be evident.

- Low boost pressure. If a turbocharger is fitted and is not supplying enough air for the fuel injected, this would be another form of air starvation.

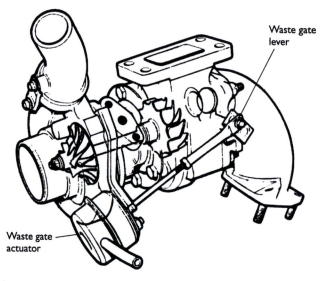

Waste gate lever

Waste gate actuator

Figure 4.32 MDi engine turbocharger

Turbocharger Figure 4.32 shows the MDi engine turbocharger, but why turbocharge at all?

- Engine power output can be increased, by up to 30%, without increasing the engine speed or displacement.
- It provides increased engine torque.
- It utilises exhaust energy to improve performance and increase fuel efficiency.
- It reduces emissions.
- It gives lower exhaust system noise.

Remember:

- A turbocharger is a component matched to a particular engine. It is not a 'bolt-on' accessory.
- A turbocharger will not correct an engine fault, but will more likely enhance it.
- Do not try to update a turbocharger by altering the boost pressure.
- The correct grade of oil must be used and the filter must be changed at recommended intervals.
- The air filter must be changed at the recommended intervals.
- A turbocharger is a finely balanced, precision component.
- Providing all the required precautions and servicing are adhered to, there is very little to go wrong with a turbocharger, and it should never be condemned without carrying out a thorough engine check first.

LEARNING TASKS

➡ Look back at the key words. Explain each one to a friend, and/or write out a short description to keep as evidence.

➡ Make a simple sketch to show the layout of the MDi engine diesel fuel system.

➡ Examine a real system, if available, and note the construction techniques and materials used.

➡ Write a short explanation about what you consider to be the advantages and disadvantages of the MDi engine.

6 BMW 328i six cylinder engine – case study

Introduction

KEY WORDS

- Torque and power curves
- On board diagnostics
- Variable camshaft control
- Knock control

The six-cylinder engine from BMW (Figure 4.33) offers improved performance on less fuel with even greater refinement. Figure 4.34 shows the torque and power curves for this engine. The following list gives some of the reasons for these improvements:

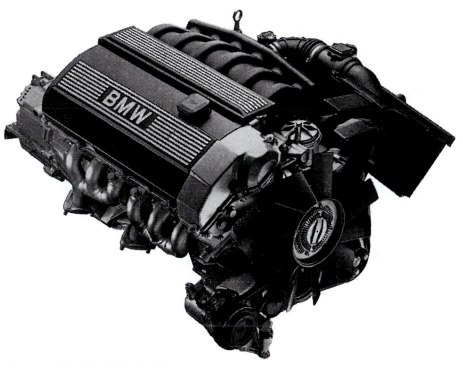

Figure 4.33 The BMW 2.8 litre six cylinder engine uses variable camshaft control

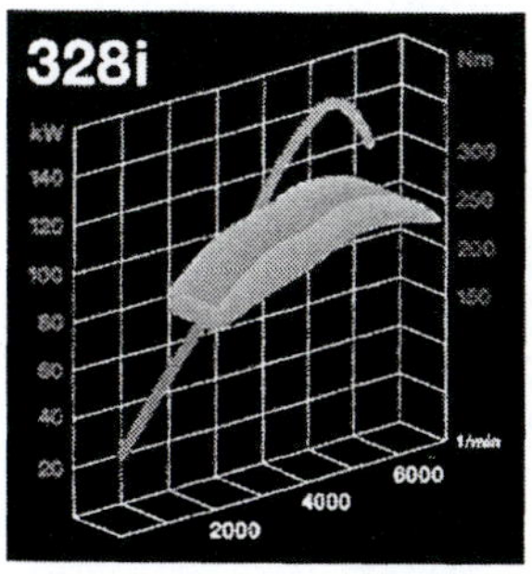

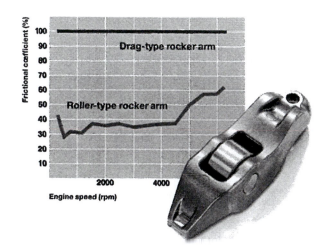

Figure 4.34 Torque and power curves for the BMW 328i with a 2.8 litre engine. Variable valve timing improves the torque curve. (Good torque through a large speed range.)

- all aluminium engine available as 2.5 or 2.8 litre options
- reduced weight exhaust system
- reduced friction in the engine (Figure 4.35)
- sophisticated Digital Motor Electronics (DME)
- four valve technology
- variable valve timing (Figure 4.36)
- on board diagnostics

Figure 4.35 Roller-type rocker arms (BMW)

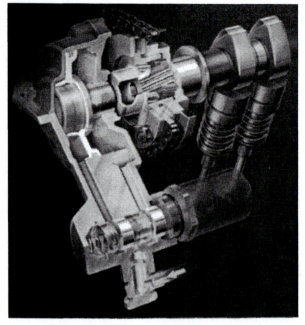

Figure 4.36 Variable valve timing from BMW

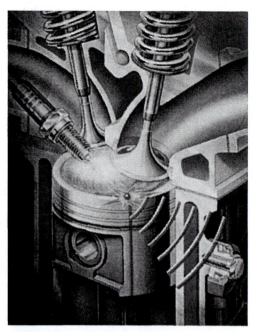

Figure 4.37 Cylinder-specific knock control gets the most out of the fuel consumed, ensuring that the engine runs constantly with maximum efficiency

- individual control intake manifold
- cylinder-specific knock control (Figure 4.37).

The following section will examine some of these issues in a little more detail.

System details A modified valve drive system uses roller type rocker arms, which reduce friction on the camshaft by up to 60%. The result of this is an improvement in fuel economy and increase in engine power. The Digital Motor Electronics, known by BMW as DME, is what we would describe as engine management (see chapter 5). This system controls ignition, injection and valve timing as well as other functions. This level of control guarantees optimum power and performance on minimum fuel and with enhanced emission control. The DME system also has on board diagnostics (OBD). Its task is to recognise faults at an early stage before they can do any harm. These faults are memorised and displayed on suitable test equipment.

Variable camshaft control is used to good effect. Data that is provided by the control unit is used to determine and then adjust the inlet camshaft by up to 12.5°. This is done hydromechanically as a function of engine speed. The inlet valves are opened later at low speeds in order to improve idling quality. At medium speeds, the inlet valves are opened earlier. Advancing the opening provides more torque and allows more exhaust gas to be recirculated. This, in turn, improves economy and reduces emissions. At higher speed, the open time of the inlet valves is advanced further still. This gives maximum power and better performance.

The individual control inlet manifold helps to boost engine torque at low speeds thanks to an increase in intake air speed. At high engine speeds, a connection pipe opens up between the intake manifolds. This shortens the inlet and produces extra power. As a general rule, long tract inlet manifolds are best at low engine speeds and shorter tracts are best at high speeds.

Cylinder-specific knock control offers maximum power and performance on minimum fuel. As soon as the engine is about to knock, sensors inform the control unit and the timing is adjusted. Most engines run best with the ignition timing advanced as far as possible. However, if advanced too far, the engine will suffer from combustion knock and would be damaged very quickly.

Most of the ideas explained here are not necessarily new, but BMW has combined them and advanced the technology to a high level with excellent results.

LEARNING TASKS

➡ Look back at the key words. Explain each one to a friend, and/or write out a short description to keep as evidence.

➡ Write a short explanation about what you consider to be the advantages and disadvantages of the BMW 328i engine.

5 Engine management systems

1 Introduction

Start here! 'Engine management' is a general term that describes the control of engine operation. This can range from a simple carburettor controlling or managing the fuel, with an ignition distributor with contact breakers to control the ignition, to a very sophisticated electronic control system.

By this stage, I'm sure you have a good idea about the basic operation of fuel and ignition systems (if not, jump back to the first book to brush up on this subject). If you have got the basics, then this chapter will be easier to follow. This is because the fundamental task of an engine management system is to further refine the basic control of an engine.

KEY WORDS

- All words in the table
- Management
- Ignition timing
- Fuel mixture control

Figure 5.1 Engine management ECU

More accurate timing and accurate fuel control are the main aims of an engine management system.

Terminology	
RAM – Random Access Memory	A digital memory like in a computer that is volatile, in other words it will 'forget' if you turn the power off. In an engine management system, it is used to store information about engine operating variables such as speed, load and temperature
ROM – Read Only Memory	Digital memories that can hold information permanently, a bit like a book. In an engine management system, it is used to hold pre-programmed data about engine requirements, as well as the computer program to make the system work
ECU – Electronic Control Unit	The heart of an electronically-operated system. Most engine management ECU's are like a computer. They take in information, decide what to do and cause some action to occur
Sensor	A device which converts a variable condition, such as engine speed, into an electrical signal that can be used by the ECU

Terminology

Actuator	A device which is able to control something or make something happen. A good example is a fuel injector
Basic control system	A very useful way to represent any complex system. The idea is to break down the complexity into inputs, control and outputs
Closed loop control	When the output of a system causes a change in the input, it is described as having closed loop control
Open loop control	When the output of a system has no effect on the input, it is described as having open loop control
Governor	A device forming part of a diesel fuel injection pump. Its purpose is to control the operation speed of an engine by controlling the fuelling
Speed droop	The ability of a governor to control engine speed is known as its speed droop. This is the increase that occurs in engine speed (as a percentage) when loading is removed from a diesel engine, but with no change in accelerator position
Pintle injector	An injector using a nozzle that, as it opens, forms a suitable shape for the injection and atomisation of fuel. Most petrol injectors are this design. Some diesel injectors work on a similar principle, when used on indirect injected diesel engines. The spray pattern forms a conical shape
Multi-hole or orifice injector	A diesel injector made so that, as a valve opens, fuel is forced through one or several very small holes. The size, position and direction of the holes are very important and will vary depending on engine design
Lambda	Lambda, represented by the Greek letter λ, is used to describe air fuel mixture. A lambda value of 1 means the ideal ratio of 14.7:1. Values greater than 1 mean a weak mixture, less than 1 a rich mixture. Modern engine systems aim to keep the lambda value between 0.97 and 1.03. This is known as the 'lambda window' and is essential for vehicles fitted with a catalytic converter

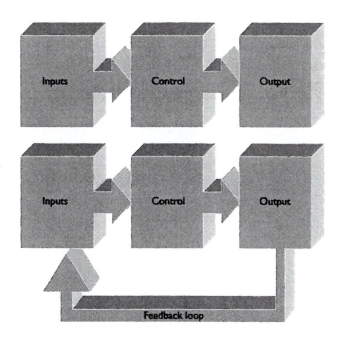

Figure 5.2 Open and closed loop control

LEARNING TASKS

➡ Look back at the key words. Explain each one to a friend, and/or write out a short description to keep as evidence.

➡ Make a simple sketch to show open and closed loop systems.

➡ Examine a real system or examine workshop manuals and note why many ~~systems are now described as having engine management~~

2 Ignition fundamentals

Introduction

The purpose of the ignition system is to supply a spark inside the cylinder, near the end of the compression stroke, to ignite the compressed charge of air/petrol mixture.

For a spark to jump across a 0.6 mm gap in an engine cylinder, a compression ratio of 8:1, about 8 kV, is required. For higher compression ratios and weaker mixtures, a voltage up to 20 kV may be required. The ignition system therefore has to change the normal battery voltage of 12 V to between 8 to 20 kV. It also has to deliver the high voltage to the right cylinder, at the right time.

Conventional ignition is the forerunner of the more advanced systems controlled by electronics. It is well worth mentioning at this stage, though, that the basic operation of most ignition systems is very similar. They *all* involve switching a coil on and off to create a high voltage.

Ignition timing

For maximum efficiency, the ignition timing should create maximum combustion pressure at about 10° ATDC. The ideal ignition timing depends on two factors: engine speed and engine load. An increase in engine speed requires the ignition timing to be advanced. A cylinder charge, of air fuel mixture, requires a certain time to burn, hence the need to ignite it earlier at higher engine speeds. A change in timing due to engine load is required as the weaker mixture used on low load conditions burns at a slower rate, therefore ignition advance is necessary.

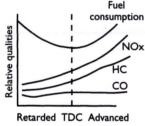

Figure 5.3 Ignition timing at a fixed speed

Spark advance is achieved in a number of ways. The simplest of these is the mechanical system comprising of a centrifugal advance mechanism and a vacuum advance unit. Manifold vacuum is almost inversely proportional to the engine load. Alternatively, I prefer to say, manifold absolute pressure is proportional to engine load.

- High engine load = high pressure = low vacuum (e.g. accelerating)
- Low engine load = low pressure = high vacuum (e.g. cruising).

Electronic ignition systems may also adjust the timing in relation to other variables such as coolant temperature and air temperature. The values of all factors that require changes in ignition timing are combined mechanically or electronically, to determine the ideal ignition point.

Energy to make a spark is stored in the ignition coil in the form of a magnetic field. To ensure the coil is charged before the ignition point, a dwell period is required. 'Dwell' is, therefore, the amount of time that is set to charge the coil; it is given as an angle, a percentage, or a time.

The ignition timing has a significant effect on fuel consumption, torque, driveability and exhaust emissions. The three most significant pollutants are hydrocarbons (HC), carbon monoxide (CO) and nitrogen oxides (NOx). The emissions of HC and NOx increase as timing is advanced. CO changes very little with timing and is mostly dependent on air fuel ratio. This makes the final timing setting a compromise.

Contact breaker ignition components

Spark plug	Seals electrodes for the spark to jump across in the cylinder. Must withstand very high voltages, pressures and temperatures
Ignition coil	Stores energy in the form of magnetism and delivers it to the distributor via the HT lead. Consists of primary and secondary windings
Ignition switch	Provides driver control of the ignition system and, usually, is also used to cause the starter to crank, and thus rotate the engine
Ballast resistor	Shorted out during the starting phase to cause a more powerful spark. Also contributes towards improving the spark at higher speeds
Contact breakers (points)	Switch the primary ignition circuit on and off to charge and discharge the coil
Capacitor (condenser)	Suppresses most of the arcing as the contact breakers open. This allows for a more rapid break of primary current and, hence, a more rapid collapse of coil magnetism, which produces a higher voltage output
HT distributor cap and rotor arm	Sends the spark from the coil to each cylinder in a pre-set sequence
Centrifugal advance	Changes the ignition timing with engine speed. As speed increases, the timing is advanced up to a pre-set value
Vacuum advance	Changes timing depending on engine load. On conventional systems, the vacuum advance is most important during cruise conditions

KEY WORDS

- All words in the tables
- ESA
- Spark plug heat range
- Distributorless
- Constant energy
- Timing map
- MAP

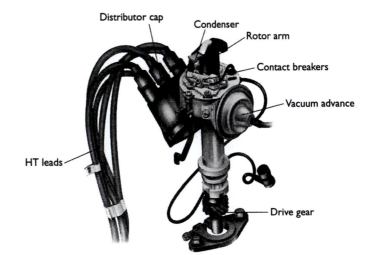

Figure 5.4 Distributor with contact breakers

Electronic ignition systems

Electronic ignition is now fitted almost universally to petrol engined vehicles. This is because the conventional system has three major disadvantages.

- Mechanical problems with the contact breakers, not least of which is the limited lifetime.

- Current flow in the primary circuit is limited to about 4A or damage will occur to the contacts – or, at least, the lifetime will be seriously reduced.

- Legislation requires stringent emission limits which means that the ignition timing must stay in tune for a long period of time.

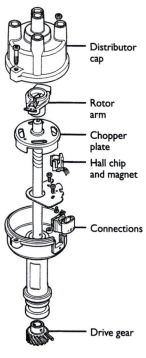

Distributor cap

Rotor arm

Chopper plate

Hall chip and magnet

Connections

Drive gear

Figure 5.5 Hall effect pulse generator in a distributor

These problems can be overcome by using a transistor to carry out the switching of the coil, and a pulse generator to tell the transistor when to switch.

Constant dwell

The term 'dwell', when applied to ignition, is a measure of the time during which the ignition coil is charging. In other words, it is the length of time the coil is switched on. The dwell in conventional systems was simply the time or angle during which the contact breakers were closed. This is now often expressed as a percentage of one charge-discharge cycle.

Whilst a very good system in its time, at very high engine speeds the reduced time available to charge the coil produced a lower power spark. As engine speed increases, dwell angle or dwell percentage remains the same, but the actual time is reduced. Constant energy solves this problem.

Constant energy

To give a constant energy ignition, the electronic ignition must increase dwell with engine speed. To provide benefit, the ignition coil must be charged up to its full capacity, in a very short time (the time of maximum dwell at the highest expected engine speed). Constant energy coils are very low resistance and are able to charge up very quickly. Typical resistances are less than 1Ω. 'Constant energy' means that, within limits, the energy available to the spark plug remains constant under all operating conditions.

Due to the high-energy operation of constant energy ignition coils, the coil cannot be allowed to remain switched on for more than a certain time, as it would overheat. This is not a problem when the engine is running, as the dwell is varied electronically from a small percentage at low speeds, to a larger percentage at high engine speed. This keeps the dwell *time*, and therefore the *energy*, stored in the coil constant. Some form of protection must also be provided for when the ignition is switched on but the engine is not running. This is known as 'stationary engine primary current cut off'.

Distributor pulse generators

Hall effect

The Hall effect distributor has become very popular with many manufacturers. Figure 5.5 shows a typical distributor with a Hall effect sensor. As the central shaft of the distributor rotates, the chopper plate attached under the rotor arm alternately covers and uncovers the Hall chip. The number of vanes corresponds with the number of cylinders. The vanes cause the Hall chip to be alternately in and out of a magnetic field. The result of this is that the device will produce almost a square wave output, which can then easily be used to switch further electronic circuits. The three terminals on the distributor are marked '+ 0 −'. The terminals, '+' and '−', are a voltage supply, and terminal '0' is the output signal. Hall effect distributors are very common due to the accurate signal produced and long term reliability. Operation of a Hall effect pulse generator can easily be tested with a DC voltmeter or a logic probe. Note tests must not be carried out using an ohmmeter as the voltage from the meter can damage the Hall chip.

Inductive

Inductive pulse generators use the basic principle of induction to produce a signal typical of the one shown in Figure 3.24. Many forms exist but they are all based around a coil of wire and a permanent magnet. The example distributor shown in Figure 5.6 has the coil of wire wound on the pick-up and, as the reluctor rotates, the magnetic flux varies due to the peaks on the reluctor. The number of peaks or teeth on the reluctor corresponds to the number of engine cylinders.

Capacitor discharge ignition

Capacitor discharge ignition, or CDI, has been in use for many years on some models of the Porsche 911 and some Ferrari models. The CDI works by first stepping up the battery voltage to about 400 V DC, using an oscillator and a transformer, followed by a rectifier. This high voltage is then used to charge a capacitor. At the point of ignition the capacitor is discharged through the primary winding of a coil. This rapid discharge through the coil primary will produce a very high voltage output from the secondary winding. This very high voltage will ensure that even a carbon- or oil-fouled plug will be fired. The disadvantage however is that the spark duration is short, which can cause problems, particularly during starting. This can be overcome, however, by using direct ignition.

Programmed ignition

'Programmed ignition' is the term used by Rover and some other manufacturers. Ford, Bosch and others call it 'electronic spark advance' or ESA. Basic constant energy electronic ignition was a major step forward and is still used on many applications. However, its limitations lay in still having to rely upon mechanical components for speed and load advance characteristics. In many cases, these did not match ideally the requirements of the engine.

Programmed ignition systems differ from earlier systems in that they operate digitally. In this system, information about the operating requirements of a particular engine can be programmed into the memory inside the electronic control unit. The data for storage in ROM, or read only memory, is obtained from rigorous testing on an engine dynamometer and further development work in the vehicle under various operating conditions.

Programmed ignition has several advantages.

■ The ignition timing can be accurately matched to the individual application under a range of operating conditions.

■ Other control inputs can be utilised, such as coolant temperature and ambient air temperature.

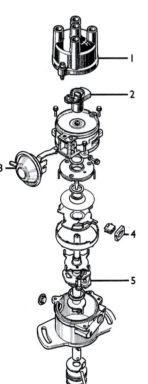

1. Distributor cap
2. Rotor arm
3. Vacuum advance
4. Connections to coil of wire
5. Centrifugal advance mechanism

Figure 5.6 Inductive pulse generator in a distributor

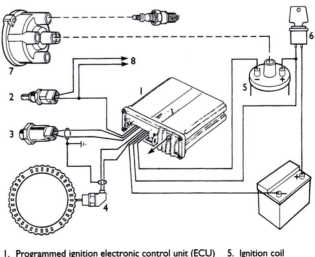

1. Programmed ignition electronic control unit (ECU)
2. Coolant temperature sensor
3. Knock sensor
4. Crankshaft sensor and reluctor
5. Ignition coil
6. Ignition switch
7. High tension distributor
8. Fuel injection ECU

Figure 5.7 Programmed ignition system

- Starting is improved, fuel consumption and emissions are reduced and idle control is better.

- Other inputs, such as engine knock, can be taken into account.

- The number of wearing components in the ignition system is considerably reduced.

Programmed ignition or ESA can be a separate system or included as part of the fuel control system. This gives limitless possibilities in managing the control of the engine.

Figure 5.7 shows the layout of a programmed ignition system. In order for the ECU to calculate suitable timing and dwell outputs, certain input information is required.

Engine speed and position: crankshaft sensor	The device consists of a permanent magnet, a winding and a soft iron core. It is mounted close to a reluctor disc. The disc has teeth spaced at intervals around the periphery of the disc. It has teeth missing at a known position before TDC. Many manufacturers use this technique with minor differences. As a tooth from the reluctor disc passes the core of the sensor, the reluctance of the magnetic circuit is changed. This induces a voltage in the winding. The frequency of the waveform being proportional, the engine speed. The missing tooth causes a 'missed' output wave and hence engine position can be determined
Engine load: manifold absolute pressure sensor	Engine load is proportional to manifold pressure in that high load conditions produce high pressure and lower load conditions, such as cruise, produce lower pressure. Load sensors are therefore pressure sensors. They are either mounted in the ECU or as a separate unit, and are connected to the inlet manifold with a pipe. The pipe often incorporates a restriction to damp out fluctuations and a vapour trap to prevent petrol fumes reaching the sensor
Engine temperature: coolant sensor	Coolant temperature measurement is carried out by a simple thermistor. The thermistor changes its resistance as temperature changes
Detonation: knock sensor	Combustion knock or detonation can cause serious damage to an engine if sustained for long periods. This knock or detonation is often caused by over advanced ignition timing. At odds with this is the fact that an engine will run at its most efficient when the timing is advanced as far as possible. The sensor is fitted in the engine block between cylinders two and three on in-line four engines. The ECU reacts to signals from the knock sensor in the engine's 'knock window' for each cylinder. This is often just a few degrees each side of TDC. If detonation is detected, the ignition timing is retarded in steps until knock is no longer detected. The steps vary between manufacturers but 2° steps are typical. The timing is then advanced slowly in steps of less than 1° over a number of engine revolutions, until the advance required by memory is restored. This fine control allows the engine to be run very close to knocking but without risk of engine damage
Battery voltage	Correction to dwell settings is required if the battery voltage falls. A lower voltage supply to the coil will require a slightly larger dwell figure

Electronic control unit

Electronic systems have become much more sophisticated, with a big increase in the information that can be held in the memory chips of the ECU. The earlier versions of programmed ignition systems produced achieved accuracy in ignition timing of ±1.8°, whereas a conventional distributor is ±8°. The information, which is derived from dynamometer tests as well as running tests in the vehicle, is stored in ROM. The basic timing map consists of the correct ignition advance for various engine speed and engine load conditions. This is shown in Figure 5.8 using a cartographic (map-like) representation.

A separate three-dimensional map is used to add corrections for engine coolant temperature to the basic timing settings. This improves driveability, and can be used to decrease the warm up time of the engine. The ECU will

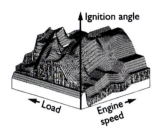

Figure 5.8 Cartographic map representing how ignition timing is stored in the ECU

also make corrections to the dwell angle, both as a function of engine speed, to provide constant energy output, and corrections due to changes in battery voltage. A lower battery voltage will require a slightly longer dwell, and a higher voltage will require a slightly shorter dwell.

The output of a system such as this programmed ignition is very simple. The output stage in common with all electronic ignition systems consists of a heavy-duty transistor. This is simply to allow the high ignition primary current to be controlled. The switch off point of the coil will control ignition timing and the switch on point will control the dwell period.

HT distribution

The high-tension distribution is similar to a more conventional system. The rotor arm, however, is mounted on the end of the camshaft, with the distributor cap positioned over the top. The distributor cap is mounted on a base plate which, as well as acting as the mounting point, prevents any oil which leaks from the camshaft seal fouling the cap and rotor arm.

Distributorless ignition

Distributorless ignition used extensively by Ford has all the features of electronic spark advance (ESA) systems but, by using a special type of ignition coil, outputs to the spark plugs without the need for an HT distributor.

The system is generally only used on four or six cylinder engines, as the control system becomes more complex for higher numbers. The basic principle is that of the 'lost spark'. The distribution of the spark is achieved by using two double-ended coils, which are fired alternately by the ECU. The timing is determined from a crankshaft speed and position sensor as well as load and other sensors. When one of the coils is fired, a spark is delivered to two engine cylinders, 1 and 4, or 2 and 3. The spark delivered to the cylinder on the compression stroke will ignite the mixture as normal. The spark produced in the other cylinder will have no effect, as this cylinder will be just completing its exhaust stroke.

Because of the low compression and the exhaust gases in the 'lost spark' cylinder, the voltage used for the spark to jump the gap is only about 3 kV. This is similar to the more conventional rotor arm to cap voltage. The spark produced in the compression cylinder is therefore not affected.

The DIS system consists of three main components: the electronic module, a crankshaft position sensor, and the DIS coil. In many systems a manifold absolute pressure sensor is integrated in the module. The module functions in much the same way as has been described for the previously described electronic spark advance system. The crankshaft position sensor is again similar in operation to the one described in the previous section.

A DIS coil is shown in Figure 5.9. The primary winding is supplied with battery voltage to a centre terminal. The appropriate half of the winding is then switched to earth in the module. The secondary windings are separate and are specific to cylinders 1 and 4, or 2 and 3.

Figure 5.9 Multiple-spark ignition coil for high-tension distribution

Direct ignition

Direct ignition is in a way the follow on from distributorless ignition. This system uses a coil for each cylinder. These coils are mounted directly on the spark plugs. Figure 5.10 shows the Saab direct ignition system. The use of an individual coil for each plug ensures an extended charge time. The low resistance primary winding is, therefore, built up to a high energy level. This ensures that a very high voltage, high-energy spark is produced. This

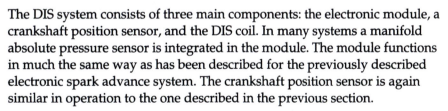

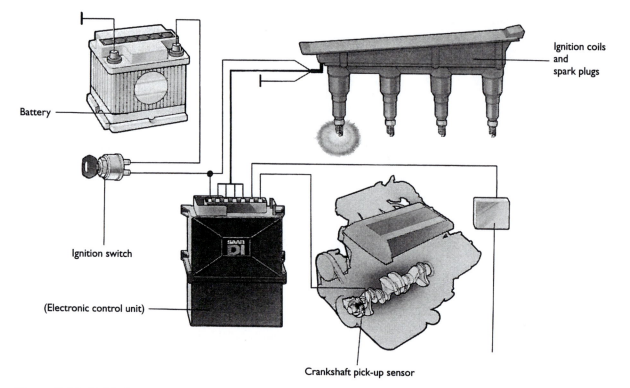

Battery

Ignition coils and spark plugs

Ignition switch

(Electronic control unit)

Crankshaft pick-up sensor

Figure 5.10 SAAB direct ignition system

voltage, which can be in excess of 40 kV, provides efficient starting of the combustion process under cold starting conditions and with weak mixtures. Some direct ignition systems use capacitor discharge ignition.

Spark plugs The function of a spark plug is to allow a spark to form within the combustion chamber, which will initiate burning. In order to do this, the plug has to withstand a number of severe conditions. Let us take as an example a four cylinder four-stroke engine with a compression ratio of 9 to 1, running at speeds up to 5000 rev/min. At this speed the four-stroke cycle will repeat every 24 ms. The following conditions are typical:

- end of induction stroke 0.9 bar at 65°C
- ignition firing point 9 bar at 350°C
- highest value during power stroke 45 bar at 3000°C
- power stroke completed 4 bar at 1100°C.

Beside the above conditions, the spark plug must withstand severe vibration and a harsh chemical environment. Finally, but perhaps most important, the insulation properties must withstand voltages up to 40 kV.

Figure 5.11 shows a typical spark plug. The centre electrode is connected to the top terminal by a stud. The electrode is constructed of a nickel-based alloy. Silver and platinum are also used for some applications. If a copper core is used in the electrode, this improves the thermal conduction properties.

The insulating material is ceramic based and of a very high grade. Aluminium oxide is a popular choice; it is bonded into the metal parts and glazed on the outside surface. Ribs prevent tracking down the outside of the plug insulation. The ribs increase the surface distance from the terminal to the metal fixing bolt. This helps to reduce tracking down the side of the plug in damp conditions.

Figure 5.11 Bosch platinum spark plug

Heat range

Due to many and varied constructional features involved in the design of an engine, the range of temperatures a spark plug is exposed to will vary significantly. The operating temperature of the centre electrode of a spark plug is critical. If the temperature becomes too high then pre-ignition may occur, as the fuel–air mixture may become ignited due to the heat of the plug electrode. On the other hand, if the electrode temperature is too low, then carbon and oil fouling can occur, because deposits are not burnt off. The ideal operating temperature of the plug electrode is between 400°C and 900°C. Figure 5.12 shows how the temperature of the electrode changes with engine power output.

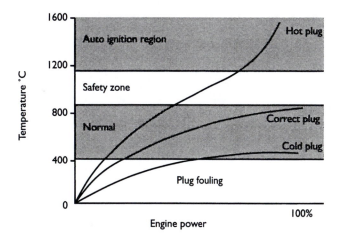
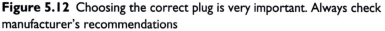

Figure 5.12 Choosing the correct plug is very important. Always check manufacturer's recommendations

The heat range of a spark plug, then, is a measure of its ability to transfer heat away from the centre electrode. A hot running engine will require plugs with higher heat transfer ability than a colder running engine. Note that hot and cold running of an engine in this sense refers to the combustion temperature and *not* to the efficiency of the cooling system.

The following factors determine the heat range of a spark plug:

- insulator nose length
- electrode material
- thread contact length
- projection of the electrode.

All these factors are dependent on each other, and the position of the plug in the engine has a particular effect.

Electrode materials

For normal applications, alloys of nickel are used for the electrode material. Chromium, manganese, silicon and magnesium are examples. These alloys exhibit excellent properties with respect to corrosion and burn off resistance. To improve on the thermal conductivity, compound electrodes are used. A common example of this type of plug is the copper core spark plug.

Silver electrodes are used for specialist applications as silver has very good thermal and electrical properties. Platinum tips are used for some spark plug applications due to the very high burn-off resistance of this material. It is also possible because of this to use much smaller diameter electrodes, thus increasing mixture accessibility. Platinum also has a catalytic effect further accelerating the combustion process. The life expectancy of modern platinum plugs is impressive. Some are said to last well over 80 000 km or 50 000 miles.

Electrode gap

Spark plug electrode gaps have increased as the power of the ignition systems driving the spark has increased. The simple relationship between plug gap and voltage required is that, as the gap increases, so must the voltage. Further, the energy available to form a spark at a fixed engine speed is constant. This means that a larger gap using higher voltage will result in a shorter duration spark. Therefore, a smaller gap will allow a longer duration spark. For cold starting an engine and for igniting weak mixtures the duration of the spark is critical. Likewise, the plug gap must be as large as possible to allow easy access for the mixture to prevent quenching of the flame. The final choice is a compromise reached through testing and development of a particular application. Plug gaps in the region of 0.7 to 1.2 mm are the norm at present.

LEARNING TASKS

➡ Look back at the key words. Explain each one to a friend, and/or write out a short description to keep as evidence.

➡ Make a simple sketch to show the inputs and outputs of a direct ignition system.

➡ Examine a real system and note the layout of the various ignition components.

➡ Write a short explanation about why it is not possible to adjust timing in the normal way on a programmed ignition system. Some systems do, however, allow a small timing change. See if you can find out how and why.

---

3   Carburation

Introduction

Figure 5.13 shows a simple fixed choke carburettor operating under various conditions. The float and needle valve assembly ensures a constant level of petrol in the float chamber. The Venturi causes an increase in air speed and hence a drop in pressure in the area of the outlet. The main jet regulates how much fuel can be forced into this intake air stream by the higher pressure now apparent in the float chamber. The basic principle is that, as more air is forced into the engine, then more fuel will be mixed into the air stream.

The problem with this system is that the amount of fuel forced into the air stream does not linearly follow the increase in air quantity, unless further compensation fuel and air jets are used.

A variable venturi carburettor, which keeps the air pressure in the venturi constant but uses a tapered needle to control the amount of fuel, is another method used to control fuel–air ratio. One version of this type of carburettor is explained in the next section on electronic control of carburation.

Stages of carburation

The basic principle of a carburettor is to mix fuel and air together in the correct ratios dependant on engine loads and temperature. Fuel flow is caused by the low air pressure around a spray outlet, and atmospheric pressure acting on the fuel in the float chamber.

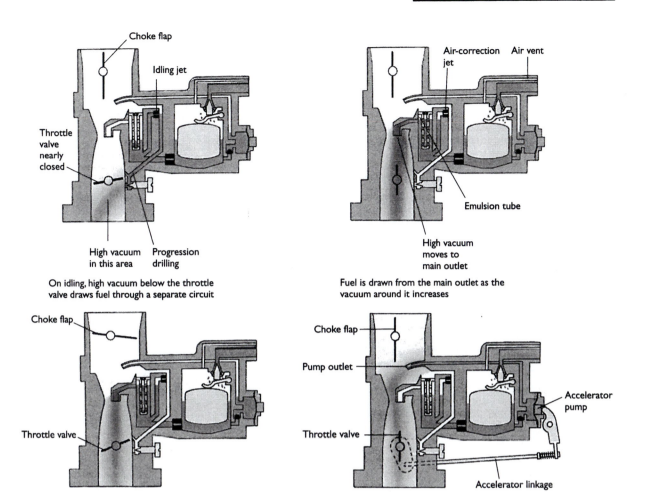

On idling, high vacuum below the throttle valve draws fuel through a separate circuit

Fuel is drawn from the main outlet as the vacuum around it increases

A flap is used to partially block the barrel for cold starts. It increases vacuum around the fuel outlet and draws more fuel to provide a rich mixture

The accelerator pump squirts an enriching shot of fuel down the barrel to provide rapid response when the throttle is opened quickly, when extra power is needed

Figure 5.13 A simple fixed choke carburettor operating under various conditions

KEY WORDS
■ All words in the table
■ Fixed choke

Whilst the engine's requirements for air–fuel mixture is infinitely variable, there are six discernible stages to consider. This can help a lot with fault finding because it helps you to 'zoom in' on the likely problem area. Although the examples given relate to a standard fixed choke carburettor, the stages are still relevant to all other types of fuel control systems.

Stage	Query or problem	Possible solution or symptoms
Cold starting – when a richer mixture is required to reduce the effects of condensation within the inlet manifold and to aid combustion	What would be the result of the choke not closing fully when starting a cold engine?	Difficult starting due to a weak mixture
	When starting a cold engine with full choke, why must the choke valve partially open when the engine is running?	To decrease the depression below the choke valve, thus preventing the mixture from becoming too rich
Idle – when the minimum amount of fuel should be provided to ensure complete combustion and efficiency	A customer complains of erratic idling and hesitation when moving off from rest?	Idling fuel jet restricted
		Mixture strength too weak at idle
	What would be the result of exchanging the idle fuel jet for one of a smaller size?	Weak idling air to fuel mixture Erratic idle and flat spots when accelerating from stationary

Stage	Query or problem	Possible solution or symptoms
Progression – when a smooth sequence of change is necessary from a range of drillings. This is normally from tickover until the main jets come into play	What is the purpose of the progression drilling?	Ensures an even change over to the main system
	What would be the result of the progression drillings becoming restricted?	Lack of response during initial acceleration with warm engine
Acceleration – when a measured increase in the supply of fuel is required to sustain an initial burst of speed	What function does the accelerator pump perform?	Enriches the mixture for hard acceleration
	The customer complains of a flat spot during hard acceleration when engine is hot?	Faulty non-return valve in the pump circuit
		Accelerator pump diaphragm holed
Cruising – where the need is for metered fuel to maintain speed at the most efficient setting	What complaint would the customer have if the main air correction jet was restricted?	Poor mid throttle range performance combined with high fuel consumption
	If mixture was too weak on cruise, what would be the effect on operation?	Possible misfiring and poor performance
		Flat spots on acceleration
High speeds – where a slightly richer mixture is required to maintain efficient combustion and to avoid damage to the engine	What operating symptoms could be caused by blocked full load enrichment tube?	Reduced top speed
	What symptoms would be observed if the mixture was too rich at high speed?	Possible poor fuel consumption, and reduced engine performance

Electronic control of carburation

Electronic control of a carburettor is made in the following areas.

Idle speed	Controlled by a stepper motor to prevent stalling but still allow a very low idle speed to improve economy and reduce emissions. Idle speed may also be changed in response to a signal from an automatic gearbox to prevent either the engine from stalling or the car from trying to creep
Fast idle	The same stepper motor as above controls fast idle in response to a signal from the engine temperature sensor during the warm-up period
Choke	A rotary choke or some other form of valve or flap operates the choke mechanism depending on engine and ambient temperature conditions
Over run fuel cut off	A small solenoid operated valve, or similar, cuts off the fuel under particular conditions. These conditions often include that the engine temperature must be above a set level, engine speed above a set level and the accelerator pedal is in the off position

The air–fuel ratio is set by the mechanical design of the carburettor, so it is very difficult to control by electrical means. Some systems have used electronic control of, say, a needle and jet, but this has not proved to be very popular.

Figure 5.14 shows the main components of the system used on some earlier vehicles. As with any control system, it can be represented as a series of inputs, a form of control and a number of outputs. This is shown as Figure 5.15.

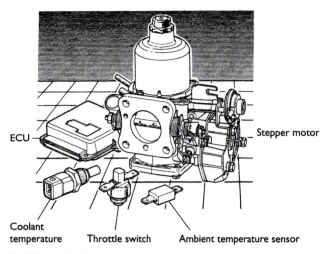

Figure 5.14 HIF variable Venturi carburettor with electronic control components

The inputs to this system are as follows:

Engine speed	This is taken from a signal wire to the negative side of the ignition coil, as is common with many systems
Engine coolant temperature	This is taken from a thermistor located in the cylinder head waterways. The same sensor is used for the programmed ignition system if fitted
Ambient temperature	A thermistor sensor is placed so as to register the air temperature. A typical position is at the rear of a headlight
Throttle switch	This switch is placed under the actual pedal and only operates when the pedal is fully off, that is, when the butterfly valve in the carburettor is closed

The main controlling actuator of this system is the stepper motor. This motor uses reduction gears to control a rotary choke valve for cold starting conditions. The same stepper motor controls idle and fast idle with a rod that works on a snail type cam. The system can operate this way because the first part of the movement of the stepper motor does not affect the choke valve; it only affects the idle speed by opening the throttle butterfly slightly. Further

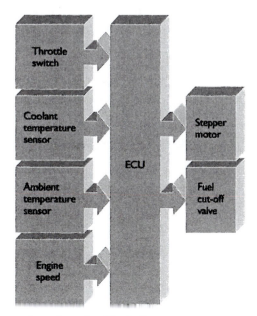

Figure 5.15 Electronic carburation block diagram

rotation then puts on the choke. The extent to which the choke is on is determined from engine temperature and ambient temperature.

The ECU 'knows' the position of the stepper motor before setting the choke position through a process known as 'indexing'. This involves the stepper motor being driven to, say, its least setting on switching off the ignition. When the ignition is next turned on, the stepper will drive the idle and choke mechanism by a certain number of steps determined, as mentioned above, by engine temperature and ambient temperature.

The other main output is the over run fuel cut off solenoid. This controls the air pressure in the float chamber and, when operated, causes pressure in the float chamber and pressure in the Venturi at the jet outlet to equalise. This prevents any fuel from being 'drawn' into the air stream.

LEARNING TASKS

➡ Look back at the key words. Explain each one to a friend, and/or write out a short description to keep as evidence.

➡ Examine a real system and note how complex some later carburettor systems are compared to basic fuel injection.

➡ Read about carburation again in the level two book to refresh your memory.

➡ Write a short explanation about how the stages of carburation could help you trace a fault.

---

4 Petrol fuel injection

Advantages of fuel injection

The basic principle of electronic fuel injection control is that, if petrol is supplied to an injector (electrically controlled valve) at a constant pressure, then the amount of fuel injected will be proportional to the injector open time.

Most systems are now electronically controlled, even if they contain some mechanical metering components. This allows the operation of the injection system to be very closely matched to the requirements of the engine. This matching process is carried out during development on test beds and dynamometers, as well as development in the car. The ideal operating data for a large number of engine operating conditions is stored in a read only memory in the ECU.

Close control of fuel quantity injected allows the optimum setting for mixture strength when all operating factors are taken into account. The major clear advantage of any type of fuel injection system is an accurate control of the fuel quantity injected into the engine.

Further advantages of electronic fuel injection include:

- over run cut off can easily be implemented
- fuel can be cut at the engines rev/min. limit
- accurate control of the fuel quantity injected into the engine allows the use of a catalytic converter
- information on fuel used can be supplied to a trip computer.

Fuel injection systems can be classified into two main categories: single point and multipoint injection. Figures 5.16 and 5.17 show the layout of these systems respectively. Both are discussed in more detail in later sections of this chapter.

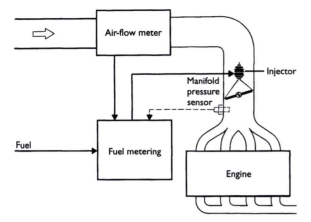

Figure 5.16 Fuel injection, single point

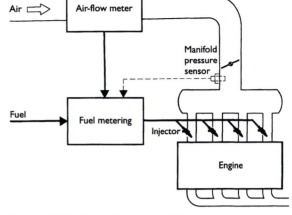

Figure 5.17 Fuel injection, multipoint

System overview

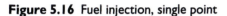

Figure 5.18 shows a typical control layout for a fuel injection system. Depending on the sophistication of the system, idle speed and idle mixture adjustment can be either mechanically or electronically controlled. Figure 5.19 shows a block diagram of inputs and outputs common to most fuel injection systems. Note that the two most important input sensors to the system are speed and load. The basic fuelling requirement is determined from these inputs in a similar way to the determination of ignition timing as described in a previous section.

Figure 5.19b shows a three-dimensional cartographic map used to represent how the information on an engine's fuelling requirements is stored. This information forms part of a read only memory (ROM) chip in the ECU.

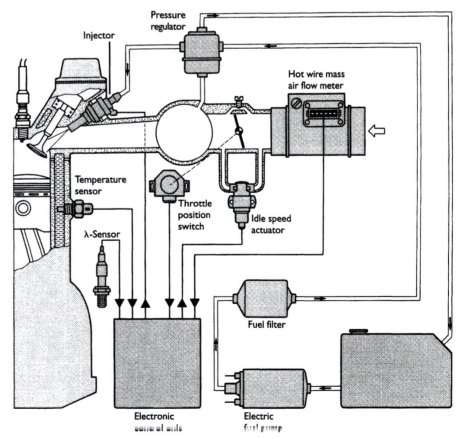

Figure 5.18 LH-Jetronic

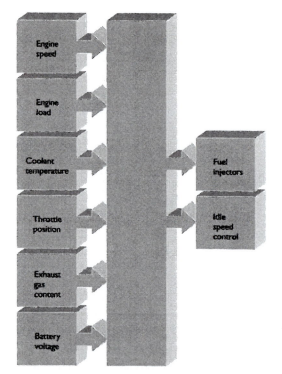

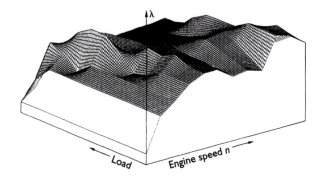

Figure 5.19b Engine fuelling (Lambda map)

Figure 5.19 Fuel injection system block diagram

When the ECU has determined the look up value of the fuel required (injector open time), corrections to this figure can be added for battery voltage, temperature, throttle change or position and fuel cut off.

Idle speed and fast idle are also generally controlled by the ECU and a suitable actuator. It is also possible to have a form of closed loop control with electronic fuel injection. This involves a lambda sensor to monitor exhaust gas oxygen content. This allows very accurate control of the mixture strength, as the oxygen content of the exhaust is proportional to the air–fuel ratio.

Components of a fuel injection system

The following parts, with some additions, are typical of the Bosch jetronic fuel injection systems. Variations from the basic system are discussed in the next section to highlight the developments.

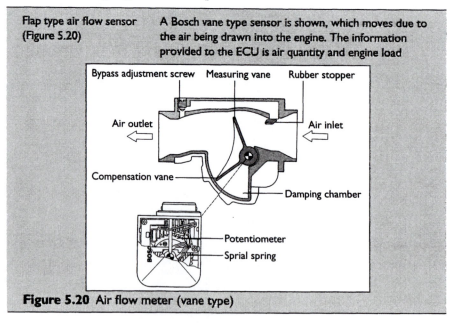

| Flap type air flow sensor (Figure 5.20) | A Bosch vane type sensor is shown, which moves due to the air being drawn into the engine. The information provided to the ECU is air quantity and engine load |

Figure 5.20 Air flow meter (vane type)

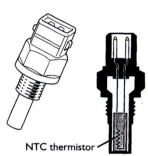

NTC thermistor

Figure 5.21 Coolant temperature sensor

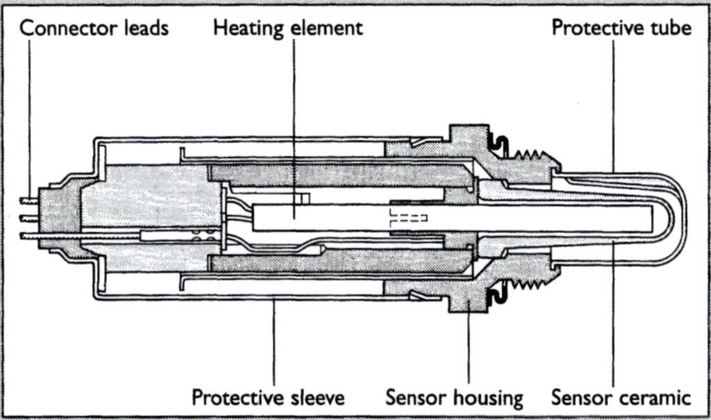

Engine speed sensor	Most injection systems, which are not combined directly with the ignition, take a signal from the coil negative terminal. This not only provides speed data but also, to some extent, engine position. A resistor in series is often used to prevent high voltage surges reaching the ECU
Temperature sensor (Figure 5.21)	A simple thermistor provides engine coolant temperature information
Throttle position sensor	The sensor can consist of two switches and only provide information that the throttle is at idle, full load or broadly anywhere in between. Alternately, a potentiometer gives more detailed throttle position information
Lambda sensor (Figure 5.22)	This device provides information to the ECU on exhaust gas oxygen content. From this, corrections can be applied to ensure the engine is kept at or very near to stoichiometry

Connector leads   Heating element   Protective tube

Protective sleeve   Sensor housing   Sensor ceramic

Figure 5.22 Lambda-sensor

Idle or fast idle control actuator (Figure 5.23)	The type shown here is a pulsed actuator. It is used to provide extra air for cold fast idle conditions. It does not spin round, but opens and closes in different open to closed ratios as required

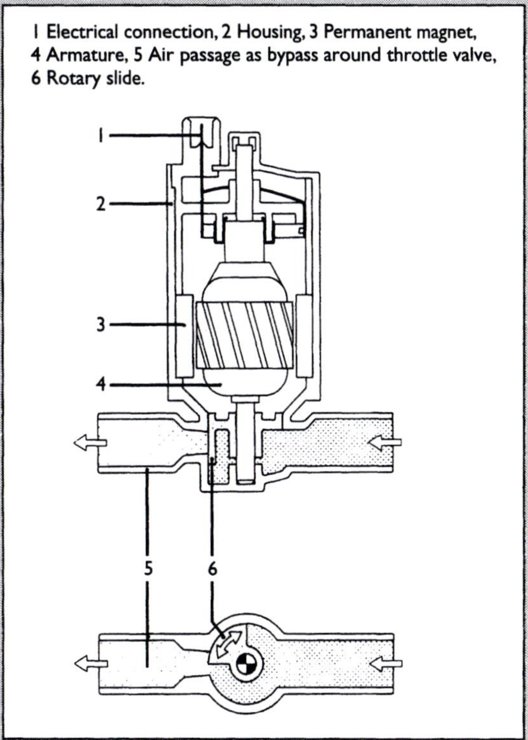

1 Electrical connection, 2 Housing, 3 Permanent magnet, 4 Armature, 5 Air passage as bypass around throttle valve, 6 Rotary slide.

Figure 5.23 Rotary idle actuator

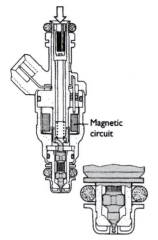

Figure 5.24 Injection valves for petrol injection systems

Fuel injector (Figure 5.24)	The pintle type is shown. It is a simple solenoid operated valve designed to operate very quickly and produce a finely atomised spray pattern
Injector resistors	These resistors were used on some systems when the injector coil resistance was very low. A lower inductive reactance in the circuit allows faster operation of the injectors. Most systems now limit injector maximum current in the ECU in much the same way as for low resistance ignition coils
Fuel pump (Figure 5.25)	The pump ensures a constant supply of fuel to the fuel rail. The volume in the rail acts as a swamp to prevent pressure fluctuations as the injectors operate. The pump must be able to maintain a pressure of about 3 bar

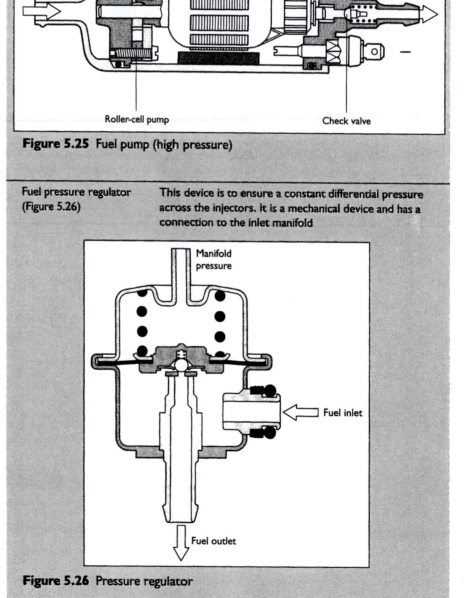

Figure 5.25 Fuel pump (high pressure)

Fuel pressure regulator (Figure 5.26)	This device is to ensure a constant differential pressure across the injectors. It is a mechanical device and has a connection to the inlet manifold

Figure 5.26 Pressure regulator

Cold start injector	An extra injector used on earlier systems as a form of choke
Thermo time switch	This works in conjunction with the cold start injector to control the amount of cold enrichment. Engine temperature and a heating winding heat it. This technique has been replaced on newer systems, which enrich by increasing the number of injector pulses or the pulse length
Combination relay	This takes many forms on different systems but is basically two relays, one to control the fuel pump and one to power the rest of the injection system. The relay is often controlled by the ECU or will only operate when ignition pulses are sensed as a safety feature. This will only allow the fuel pump to operate when the engine is being cranked or is running
Electronic control unit	Earlier ECU's were analogue in operation. All ECU's now use digital processing. This is described further in other sections

Sequential multipoint injection

Most of the systems discussed in this chapter either inject the fuel in continuous pulses, as in the single point system, or all of the multipoint injectors fire simultaneously, often injecting half of the required fuel. A sequential injection system injects fuel on the induction stroke of each cylinder in the engine firing order. This system, while more complicated, allows the stratification or layering of the cylinder charge to be controlled to some extent. This allows an overall weaker charge. Sequential injection is normally incorporated with full engine management, which is examined in a later section.

Summary

The development of fuel injection in general, and the reduced complexity of single point systems in particular, have now started to make the carburettor obsolete. As emission regulations continue to become more stringent, manufacturers are being forced into using fuel injection, even on lower-priced models. This larger market, in turn, will pull the price of the systems down, making them comparable to carburation techniques on price but superior in performance. Figure 5.27 shows the system of a mechanical fuel

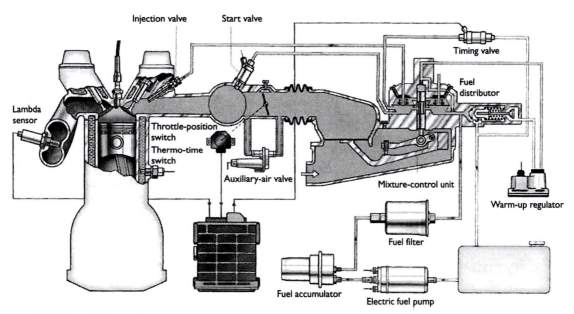

Figure 5.27 The KE-Jetronic system

injection system known as 'KE' Jetronic. Although an excellent system in its time, it is now becoming less popular due to the increased use of digital control engine management systems. Fuel injection is here to stay, even with the apparent complexities of some systems.

LEARNING TASKS

➡ Look back at the key words. Explain each one to a friend, and/or write out a short description to keep as evidence.

➡ Make a simple sketch to show the inputs and outputs of a basic fuel injection system.

➡ Examine a real system and note the type of components used.

➡ Write a short explanation about why fuel injection systems are replacing carburettor systems.

---

5   Lucas hot-wire multipoint injection – case study

Introduction   The Lucas hot-wire fuel injection system is a multipoint, indirect and intermittent injection system. In common with many other systems, the basic fuelling requirements are determined from engine speed and rate of airflow. Engine temperature and air temperature are the main correction factors. The calculation for the fuel injection period is a digital process, and the look up values are stored in a memory chip in the ECU. It is important with this and other systems that no unmetered air enters the engine except via the idle mixture screw on the throttle body. Figure 5.28 is the schematic arrangement of the hot wire system showing all the major components.

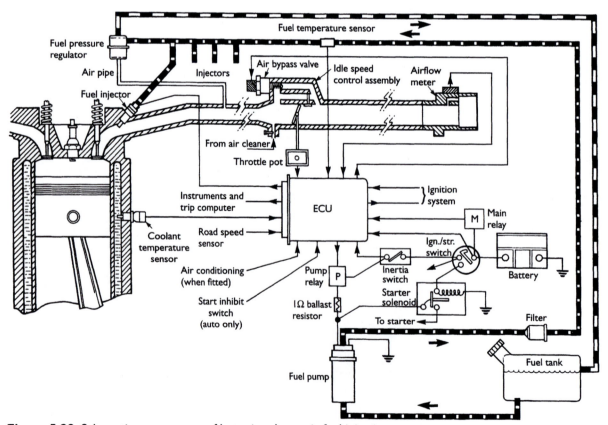

Figure 5.28 Schematic arrangement of hot-wire electronic fuel injection system

System operation   Figure 5.29 shows the components of a hot-wire fuel injection system. The ECU acts on the signals received from sensors and adjusts the length of pulse supplied to the injectors. The ECU also controls the time at which the injector pulses occur, relative to signals from the coil negative terminal. During normal running conditions, the injectors on a four-cylinder engine are all fired at the same time, and inject half of the required amount, twice during the complete engine cycle.

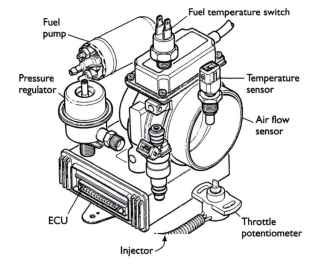

Figure 5.29 Hot wire injection system components

The fuel tank contains a swirl pot as part of the pick up pipe. This is to ensure that the pick up pipe is covered in fuel at all times, preventing air being drawn into the fuel lines. A permanent magnet electric motor is used for the fuel pump, which incorporates a roller cell type pumping assembly. An eccentric rotor on the motor shaft has metal rollers in cut outs round its edge. These rollers are forced out by centrifugal force as the motor rotates. This traps the fuel and forces it out of the pressure side of the system. The motor is always filled with fuel and the pump is able to self-prime. A non-return valve and a pressure relief valve are fitted. These will cause a pressure to be held in the system and prevent excessive pressure build up respectively.

The pump is controlled by the ECU via a relay. When the ignition is first switched on, the pump runs for a short time to ensure the system is at the correct pressure. The pump will then only run when the engine is being cranked or is running. A 1Ω ballast resistor is often fitted in the supply to the pump. This will cut down on noise but is also by-passed when the engine is being cranked to ensure the pump runs at a 'normal' speed, even when cranking causes the battery voltage to drop. An inertia switch that is usually located in the passenger compartment or boot cuts the supply to the fuel pump in the case of a collision. This is a safety feature to prevent fuel spillage. The switch can be reset by hand.

The fuel pressure across the injector must be constant if the fuel quantity injected is to be a function of the injection pulse length. This fuel pressure, which is in the region of 2.5 bar, is the difference between absolute fuel pressure and manifold absolute pressure. The fuel pressure regulator is a simple pressure relief valve with a diaphragm and spring on which the fuel pressure acts. When the pressure exceeds the pre-set value (of the spring), a valve is opened and the excess fuel returns down a pipe to the tank. The chamber above the diaphragm is connected to the inlet manifold via a pipe. As manifold pressure falls, less fuel pressure is required to overcome the spring and so the fuel pressure drops by the same amount as the manifold

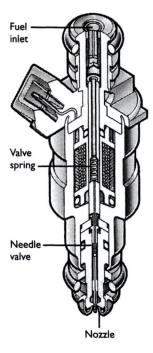

Fuel inlet

Valve spring

Needle valve

Nozzle

Figure 5.30 Fuel injector

pressure has dropped. The pressure regulator is a sealed unit and no adjustment is possible. The regulator keeps the injector differential pressure constant. This ensures that the fuel injected is only dependent on injector open time.

Figure 5.30 shows the type of injector use by this system. One injector is used for each cylinder, with each clamped between the fuel rail and the inlet manifold. The injector winding is either 2.5 or 16Ω, depending on the particular system and number of cylinders. The injectors are the needle types.

The hot wire air flow meter is the most important sensor in the system. It provides information to the ECU on air mass flow. It consists of a cast alloy body with an electronic module on the top. Air drawn into the engine passes through the main opening, with a small proportion drawn through a by-pass in which two small wires are fixed. These two wires are a sensing wire and a compensation wire. The compensation wire reacts only to the air temperature. The sensing wire is heated with a small current from the module. The quantity of air drawn over this wire will cause a cooling effect and alter its resistance, which is sensed by the module. The air flow meter has just three wires, a positive and negative supply and an output that varies between about 0 V and 5 V depending on air mass flow rate. This system can react very quickly to changes and also automatically compensates for changes in altitude. Each air flow meter is matched to its module, so repair is not normally possible.

A throttle potentiometer provides the ECU with information on throttle position and rate of change of throttle position. The device is a simple three wire variable resistor using a carbon track; it is attached to the main throttle butterfly spindle. A stable supply of 5 V allows a variable output voltage depending on throttle position. At idle the output should be 325 mV and at full load 4.8 V. The rate of change indicates the extent of acceleration or deceleration. This is used to enrich the mixture or implement over run fuel cut off as may be appropriate.

The throttle body is an alloy casting bolted to the inlet manifold and connected to the airflow sensor by a flexible trunking. This assembly contains the throttle butterfly and potentiometer, and also includes the stepper motor which controls the air by-pass circuit. Heater pipes and breather pipes are also connected to the throttle body.

The stepper motor is a four terminal, two-coil permanent magnet motor. It is controlled by the ECU to regulate idle speed and fast idle speed during the warm up period. The valve is located in an airway, which bypasses the throttle valve. The rotary action of the stepper motor acts on a screw thread. This causes the cone section at the head of the valve to move linearly, progressively opening or closing an aperture. An idle mixture screw is also incorporated in the throttle body, which allows a small amount of air to by-pass the airflow sensor.

The coolant sensor is a simple thermistor and provides information on engine temperature. The fuel temperature sensor is a switch on earlier vehicles and a thermistor on later models. The information provided allows the ECU to determine when hot start enrichment is required. This is to counteract the effects of fuel evaporation.

The heart of the system is the electronic control unit (ECU). It contains a map of the ideal fuel settings for sixteen engine speeds and eight engine loads. The figure from the memory map is the basic injector pulse width. Corrections are then added for a number of factors, the most important being the engine temperature and throttle position. Corrections are also added for some or all of the factors in the following table when appropriate.

Voltage correction	Pulse length is increased if battery voltage falls. This is to compensate for the slower reaction time of the injectors
Cranking enrichment	The injectors are fired every ignition pulse instead of every other pulse for cranking enrichment
After start enrichment	This is to ensure smooth running after starting. This is provided at all engine temperatures and decays over a set time. It is, however, kept up for a longer period at lower temperatures. The ECU increases the pulse length to achieve this enrichment
Hot start enrichment	A short period of extra enrichment, which decays gradually, is used to assist with hot starting
Acceleration enrichment	When the ECU detects a rising voltage from the throttle sensor, the pulse length is increased to achieve a smoother response. The extra fuel is needed as the rapid throttle opening causes a sudden inrush of air and, without extra fuel, a weak mixture would cause a flat spot
Deceleration weakening	The ECU detects this condition from a falling throttle potentiometer voltage. The pulse length is shortened to reduce fuel consumption and exhaust emissions
Full load enrichment	Again, this is an increase in pulse length but by a fixed percentage of the look up and corrected value
Over run fuel cut off	This is an economy and emissions measure. The injectors do not operate at all during this condition. This situation will only occur with a warm engine, throttle in the closed position, and the engine speed above a set level. If the throttle is pressed, or the engine falls below the threshold speed, the fuel is reinstated gradually to ensure smooth take up
Over speed fuel cut off	To prevent the engine from being damaged by excess speed, the ECU can switch off the injectors above a set figure. The injectors are reinstated once engine speed falls below the threshold figure

Hot wire fuel injection is a very adaptable system and will remain current in various forms for some time.

➡ Look back at the key words. Explain each one to a friend, and/or write out a short description to keep as evidence.

➡ Examine a real system or components, if available, to become familiar with the system operation.

➡ Write a short explanation about the claimed advantages of hot wire fuel injection.

6 Bosch Mono-jetronic single point injection – case study

Introduction   Single point or throttle body injection systems are very popular. This is because they have the fine control advantages of electronic injection, but are also reasonably economical to produce. Most of the reasons for manufacturers making changes to engine management systems is to comply with laws and regulations. The single point system is one solution to this requirement. The better fuel control makes the use of catalytic converters possible. However, other solutions are being developed all the time. Multipoint injection and direct petrol injection are being developed for the smaller car market.

System operation

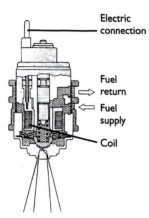

Electric connection

Fuel return

Fuel supply

Coil

Figure 5.31 Low pressure injector (mini-injector)

Mono-jetronic is an electronically controlled system with just one injector positioned above the throttle butterfly valve. The throttle body assembly looks like a carburettor. A low-pressure fuel supply (1 bar) is used to supply the injector which injects the fuel intermittently into the inlet manifold. In common with most systems, sensors measuring engine variables supply the operating data. The ECU computes the ideal fuel requirements and outputs to the injector. The width of the injector pulses determines the quantity of fuel introduced.

The injector for the TBI system is a very fast acting valve. Figure 5.31 shows the injector in section. A pintle on the needle valve is used and a conical spray patter produced. This ensures excellent fuel atomisation and hence a better 'burn' in the cylinder. To accurately meter the small fuel quantities, the valve needle and armature have a very small mass. This permits opening and closing times of much less than one millisecond. The fuel supply to the injector is continuous, this preventing air locks and ensuring a constant supply of cool fuel. This provides benefits for good hot starting performance, which can be inhibited by evaporation if the fuel is hot.

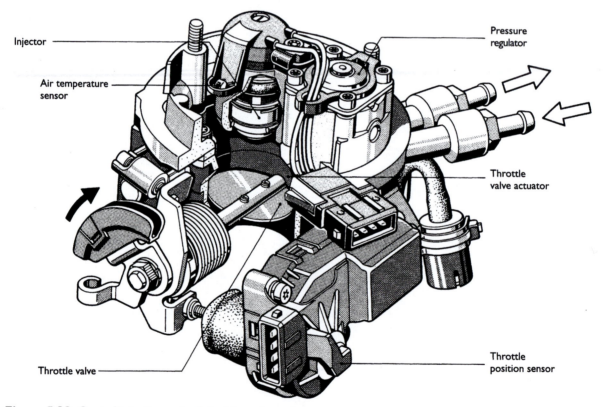

Injector

Air temperature sensor

Pressure regulator

Throttle valve actuator

Throttle position sensor

Throttle valve

Figure 5.32 Central injection unit of the Mono-jetronic

Figure 5.32 shows the main central injection unit of the mono jetronic system. You will notice that the component not used in this system is an airflow sensor. Air mass and load are calculated from the throttle position sensor, engine speed and air intake temperature. This is sometimes known as the 'speed density method'. At a known engine speed with a known throttle opening, the engine will 'consume' a known volume of air. If the air temperature is known, then the air mass can be calculated.

15 x 15

The basic injection quantity is generated in the ECU as a function of engine speed and throttle position. A ROM chip, represented by a cartographic map, stores data at 15 speed and 15 throttle angle positions, giving 225 references altogether. If the ECU detects deviations from the ideal air fuel ratio by signals from the lambda sensor, then corrections are made. If these corrections are required over an extended period, then the new corrected values are stored in memory. These are continuously updated over the life of the system. Further corrections are added to this look up value for temperature, full load and idle conditions. Over run fuel cut off and high engine speed cut off are also implemented when required.

The Bosch Mono-jetronic system also offers adaptive idle control. This is to allow the lowest possible smooth idle speed to reduce fuel consumption and exhaust emissions. A throttle valve actuator changes the position of the valve in response to a set speed calculated in the ECU, which takes into account the engine temperature and electrical loads on the alternator. The required throttle angle is computed and placed in memory. The adaptation capability of this system allows for engine drift during its life. It can also make corrections for altitude.

The electronic control unit checks that all signals fall within set limits during normal operation. If a signal deviates from the normal, this fault condition is memorised and can be output to a diagnostic tester or read as a blink code from a fault lamp.

KEY WORDS	LEARNING TASKS
■ All words in the table	➡ Look back at the key words. Explain each one to a friend, and/or write out a short description to keep as evidence.
	➡ Examine a real system or components, if available, to become familiar with the system operation.
	➡ Write a short explanation about the claimed advantages of single point fuel injection.

7 Bosch Motronic M5 full engine management – case study

Introduction The information and, therefore, the signals from sensors required to control ignition are very much the same as those required to control fuel. The ignition point and quantity of fuel injected are interrelated when it comes to issues of fuel economy, emissions and performance. This is the reason why most management systems now in use combine ignition and fuel control. The ignition and multipoint injection control system described in this section is manifold pressure sensed and uses distributorless ignition.

An electronic control unit controls this engine management system. The ECU is responsible for managing both the ignition and fuelling requirements of the engine. The system is known as multipoint simply because each cylinder is fed by its own injector. All of the injectors fire together and hence inject at the same time. The ECU uses the information provided by the sensors to ensure the engine is operated at its optimum settings. It controls two principle functions:

- the open period of the injectors
- the ignition timing.

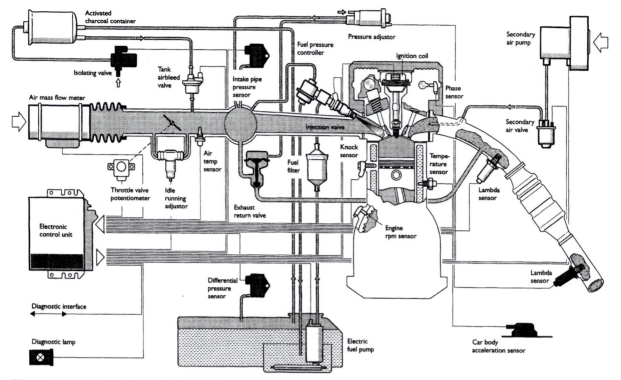

Figure 5.33 Motronic M5 with OBD II

The ECU also provides an output to the diagnostic system in the form of both a warning light and serial diagnostic link. Figure 5.33 shows the layout of the Motronic M5 system.

The combination of ignition and injection control has several advantages. The information received from various sensors is used for computing both fuelling and ignition requirements. Perhaps, more importantly, there is the fact that ignition and injection are closely linked. The influence they have on each other can be easily taken into account to ensure that the engine is working at its optimum, under all operation conditions.

Overall, this type of system is less complicated than separate fuel and ignition systems, and in many cases the ECU is able to work in an emergency mode by substituting missing information from sensors with pre-programmed values. This will allow limited but continued operation in the event of certain system failures. The system has built-in on-board diagnostics (OBD). Figure 5.34 shows some of the system components.

System operation – ignition

This is an integral ignition system without an HT distributor. The ignition process is controlled digitally by the ECU. The data for the ideal characteristics are stored in ROM from information gathered during both prototyping and development of the engine. The main parameters for ignition advance are engine speed and load, but greater accuracy can be achieved by taking further data such as engine temperature into account. This provides both optimum output and close control of pollution levels. Performance *and* pollution level control means that the actual ignition point must be a trade off between the two.

The main reference point for the ignition system is a crankshaft position sensor. This is a magnetic inductive pick up sensor positioned opposite a flywheel ring containing 58 teeth. Each tooth takes up a 6° angle of the flywheel with a 12° gap. The gap is positioned 114° BTDC for number one cylinder. Typical resistance of the sensor coil is 360Ω. The air gap between

Figure 5.34 Bosch Motronic system components

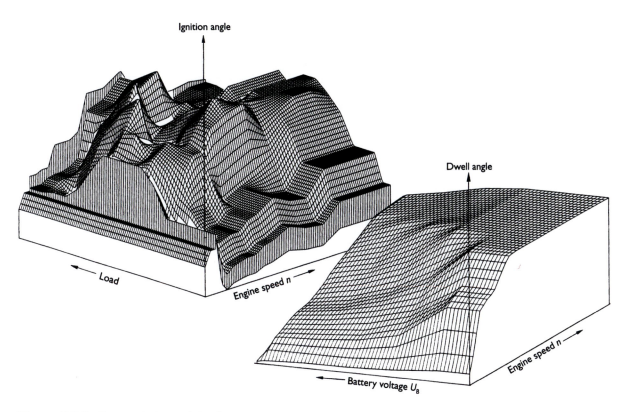

Figure 5.34b Engine timing and dwell maps

sensor and flywheel ring is about 1 mm. The signal produced is almost a sine wave but with two cycles missing corresponding to the gap in the teeth of the reluctor plate. The information provided to the ECU is:

- engine speed from the frequency of the signal
- engine position from the missing pulses.

At ignition system level, the ECU must be able to:

- determine and create advance curves
- establish constant energy
- transmit the ignition signal direct to the ignition coil.

The basic ignition advance angle, or timing, is obtained from a look up table. This is held in a ROM chip within the ECU. The look up co-ordinates for this are:

- engine rev/min – given by the flywheel sensor
- engine load – given by the air flow sensor.

Speed and load give the basic setting, but to ensure optimum advance angle the timing is corrected by:

- coolant temperature
- air temperature
- throttle butterfly position.

The electronic control unit has to determine the basic timing for a number of different conditions:

- The ignition is set to a predetermined advance during starting.
- Under idling conditions, ignition timing is moved very quickly by the ECU in order to control idle speed. When timing is advanced, engine speed will increase within certain limits.
- Full load conditions require careful control of ignition timing to prevent combustion knock. When a full load signal is sensed by the ECU (high manifold pressure), the ignition advance angle is reduced.
- The ECU will also control ignition timing variation during over run fuel cut off and reinstatement, and also to ensure anti-jerk control.
- Phase correction is when the ECU adjusts the timing to take into account the time taken for the HT pulse to reach the spark plugs.

Control of dwell angle

Figure 5.34b shows a dwell map of engine speed against battery voltage, together with a timing map. In order to maintain constant energy HT, the dwell period must increase in line with engine speed. To ensure that the ignition primary current reaches its maximum at the point of ignition, the ECU controls the dwell by use of another memory map, which also takes battery voltage into account.

The signal from the flywheel sensor is virtually a sine wave created as the teeth pass the winding. The zero value of this signal occurs as the sensor 'sees' the apex of each tooth. A circuit within the ECU converts the signal into a square wave. The passage of the missing teeth gives a longer duration signal. The ECU detects the gap in the teeth and, from this, can determine the first TDC. The second TDC in the cycle is determined by counting 29 teeth – half a revolution. The ECU, having determined the ignition angle, then controls the ignition coil every half-engine revolution. Using the reference signal, the ECU can then trigger the ignition when the number of teeth passing under the sensor corresponds to the angular movement of the crankshaft after which the spark must occur.

The ignition coil as shown in Figure 5.34, is made up of two primary windings and two secondary windings. The primary windings have a common 12 V supply and are switched to earth in turn in the normal manner. The primary resistance is in the order of 0.5Ω and the secondary resistance 14.5 KΩ. The system works on the lost spark principal, in that cylinders 1 and 4 fire together, as do 2 and 3. The disadvantage of this system is that one cylinder of each pair has the spark jumping from the plug earth electrode to the centre. It is claimed that, due to the very high energy available for the spark, this has no significant effect on performance.

Figure 5.33 shows the Motronic system using direct ignition coils as discussed in a previous section.

System operation – injection

The multipoint fuel injection system sets the amount of fuel delivered. The fuel quantity required is determined by the amount of air taken into the engine.

The ECU controls fuel supply into the engine by setting the open period of the injectors. Very accurate control of the air–fuel mixture under all operating conditions of the engine is therefore achieved. The data for the injector open period is stored in ROM in the same way as for the ignition.

Fuel is collected from the tank by a pump either immersed in it or outside but near the tank. The immersed type is quieter in operation, has better cooling and no internal leaks. The fuel is directed forwards to the fuel rail or manifold, via a paper filter.

Fuel pressure is maintained at about 2.5 bar *above manifold pressure* by a regulator mounted on the fuel rail. Excess fuel is returned to the tank. The fuel is usually picked up via a swirl pot in the tank to prevent aeration. Each of the four inlet manifold tracts has its own injector.

The fuel pump is a high-pressure type and a two-stage device. A low-pressure stage is created by a turbine drawing fuel from the tank, and a high-pressure stage is created by a gear pump delivering fuel to the filter. It is powered by a 12 V supply from the fuel pump relay, which is controlled by the ECU as a safety measure.

The fuel pump characteristics are:

- delivery – 120 lt/hour at 3 bars
- power (average consumption) – About 50 W
- resistance – 0.8Ω (static)
- voltage – 12 V
- current – 10.5 A.

The filter is placed between the fuel pump and the fuel rail. It is fitted one way only to ensure the outlet screen will trap any paper particles from the filter element. The filter will stop contamination down to between 8 and 10 microns. Replacement varies between manufacturers, but 80 000 km is often recommended.

In addition to providing a uniform supply to the injectors, the fuel rail acts as an accumulator. Depending on the size of the fuel rail, some systems also use an extra accumulator. The volume of the fuel rail is large enough to act as a pressure fluctuation damper, ensuring that all injectors are supplied with fuel at a constant pressure even as other injectors are operating.

One injector is used for each cylinder. The injectors are connected to the fuel rail by a rubber seal. In its simplest form, the injector is an electrically operated valve manufactured to a very high precision. The injector comprises of a body and needle attached to a magnetic core. When the winding in the injector housing is energised, the core or armature is attracted and the valve opens, compressing a return spring. The fuel is delivered in a fine spray to wait behind the closed inlet valve until the induction stroke begins. Providing the *pressure across the injector* remains constant, the quantity of fuel admitted is related to the open period. The open period is determined by the time the electromagnetic circuit is energised. Typically, the injectors have the following characteristics:

- supply voltage – 12 V
- resistance – 16Ω
- static output – 150 cc per minute at 3 bars.

The purpose of the fuel pressure regulator is to maintain differential pressure across the injectors at a pre-determined value. This means the regulator must adjust the fuel pressure in response to changes in manifold pressure. It is made of two compressed cases containing a diaphragm, a spring and a valve.

The spring tension and changes in manifold pressure determine the setting of the regulator. When the fuel pressure is sufficient to move the diaphragm, the valve opens and allows fuel to return to the tank. The decrease in pressure in the manifold, also acting on the diaphragm, at say, idle speed, will allow the valve to open more easily, maintaining a constant differential pressure between the fuel rail and the inlet manifold. This is a constant across the injectors and, hence, the quantity of fuel injected is determined only by the open time of the injectors. The differential pressure is maintained at about 2.5 bar.

The air supply circuit will vary considerably between manufacturers. An individual manifold from a collector housing, into which the air is fed via a simple butterfly, supplies each cylinder. The air is supplied from a suitable filter. A supplementary air circuit is used during the warm up period after a cold start.

The quantity of fuel to be injected is determined primarily by the quantity of air drawn into the engine and the load of the engine. This is dependent on two factors:

- engine rev/min.
- inlet manifold pressure.

This speed load characteristic is held in the ECU memory in ROM and can be represented by the usual cartographic map. A sensor connected to the manifold by a pipe senses manifold pressure. It is of the piezo-electric type, where the resistance varies with pressure. The sensor is fed with a stabilised 5 V supply and transmits an output voltage according to the pressure. The sensor is housed within the ECU and a pipe is required to connect the ECU to the inlet manifold. An extra 'volumetric capacity' is usually fitted in this line to damp down pressure fluctuations. The output signal varies between about 0.25 V at 0.17 bar to about 4.75 V at 1.05 bar.

The density of air varies with temperature such that the information from the MAP sensor on air quantity would be incorrect over wide temperature variations. An air temperature sensor is used to inform the ECU of the inlet air temperature such that the ECU may correct the quantity of fuel injected.

As the temperature of air decreases its density increases and so the quantity of fuel injected must also be increased. The temperature sensor is a negative temperature coefficient resistor (NTC). The resistance value decreases as temperature increases.

In order to operate the injectors, the ECU needs to know the injection point as well as, the required open time. The same flywheel sensor used by the ignition system provides this information. All four injectors operate simultaneously, once per engine revolution, injecting half of the required fuel. This helps to ensure balanced combustion. The start of injection varies according to ignition timing.

A basic open period for the injectors is determined by using the ROM information relating to:

- engine load
- engine speed.

Two corrections are then made relative to:

- temperature
- engine idling, full or partial load.

The ECU then executes another group of corrections if applicable:

- after start enrichment
- operational enrichment
- acceleration enrichment
- weakening on deceleration
- cut off on over run
- reinstatement of injection after cut off
- finally, the ECU modifies the injector open time to correct for battery voltage variation.

Under starting conditions, the injection period is calculated differently. This is determined mostly from a set figure varied as a function of temperature.

The coolant temperature sensor is a thermistor and is used to provide a signal to the ECU relating to engine coolant temperature. The ECU can then calculate any corrections to fuel injection and ignition timing. The operation of this sensor is the same as for the air temperature sensor.

The throttle potentiometer is fixed on the throttle butterfly spindle and informs the ECU of throttle position and rate of change of throttle position. The sensor provides information on acceleration, deceleration and whether the throttle is in the full load or idle position. It comprises of a variable resistance and a fixed resistance. In common with many sensors, a fixed supply of 5 V is provided and the return signal will vary approximately between 0 V and 5 V. The voltage increases as the throttle is opened.

Phases of operation

The operation functions employed by this system can be examined under a number of headings or phases as follows:

Starting phase	Entry to the starting phase occurs as soon as the ECU receives a signal from the flywheel sensor. The ignition advance is determined relative to the engine speed and the water temperature. The ECU operates the injectors four times per engine cycle (twice per crankshaft revolution) in order to obtain the most uniform mixture and to avoid wetting the plugs during the starting phase. Injection ceases 24° after the flywheel TDC signal. The ECU sets an appropriate injection period corrected in relation to water temperature if starting from cold, and air temperature if starting from hot. Exit from this starting phase occurs when the engine speed passes a threshold determined by water temperature
After start enrichment phase	Enrichment is necessary to avoid stalling after starting. The amount of enrichment is determined by the ECU, acting on information from air and coolant temperature sensors. If the engine is cold or an intermediate temperature, the initial mixture is a function of water temperature. If the engine is hot, the initial mixture is a function of air and water temperature. If the engine happens to stop within a certain period of time just after a cold start, the next post start enrichment will be reduced slightly
Engine cold running phase	During warm up, the ignition timing is corrected in relation to water temperature. Timing will also alter depending on engine speed and load. During the warm up phase, the injector open period is increased by the coolant temperature signal to make up for fuel losses and to prevent the engine speed dropping. The enrichment factor is reduced as the resistance of the temperature sensor falls, finally ceasing at 80°C. The enrichment factor is determined by engine speed and temperature at idle, and at other times by the programmed injection period relative to engine speed as well as the water temperature. To overcome the frictional resistances of a cold engine, it is important to increase the mixture supply. This is achieved by using an idle running adjuster, which allows air to bypass the throttle butterfly. This will influence the manifold pressure sensor, and the fuel will be increased accordingly. The device controls air bypass by pulsing movements. The on/off ratio determines how much air is allowed through
Idling phase	Air required for idling bypasses the throttle butterfly by a passage in the throttle housing. A volume screw is fitted for adjustment of idle speed. Idle mixture adjustment is carried out electronically in response to the adjusting of a potentiometer either on the ECU or as a separate unit. The ignition and injection functions for idle condition are set in response to information from the throttle potentiometer that the throttle is at idle position and engine speed from the flywheel sensor
Full load phase	Under full load conditions, the ignition timing is related to engine speed and full load information from the throttle potentiometer. The injection function, in order to achieve maximum power, must be set such that the mixture ratio is increased to 12.5:1. The information from the throttle potentiometer triggers a programme in the ECU to enrich the mixture in relation to engine speed, in order to ensure maximum power over the

	speed range but also to minimise the risk of knocking. It is also important not to increase fuel consumption unnecessarily, and not to allow significant increases in exhaust emissions
Acceleration phase	When a rapid acceleration is detected by the ECU, from the rate of change of the throttle potentiometer signal, enrichment occurs over a certain number of ignitions. The enrichment value is determined from water temperature and pressure variations in the inlet manifold. The enrichment then decreases over a certain number of ignitions. The enrichment is applied for the calibrated number of ignitions and then reduced at a fixed rate until it is non-existent. Acceleration enrichment will not occur if the engine speed is above 5000 rev/min or at idle
Deceleration phase	If the change in manifold pressure is greater than about 30 mbar, the ECU causes the mixture to be weakened relative to the detected pressure change
Injection cut off on deceleration phase	This is designed to improve fuel economy and to reduce particular emissions of hydrocarbons. It will occur when the throttle is closed and the engine speed is above a threshold related to water temperature (about 1500 rev/min.). When the engine speed falls to about 1000 rev/min., injection recommences with the period rising to the value associated with the current engine speed and load
Knock protection phase	Ignition timing is also controlled to reduce jerking and possible knocking during cut off and reinstatement. The calculated advance is reduced to keep the ignition just under the knock limit. The advance correction against knock is a programme relating to injection period, engine speed and water/air temperature
Engine speed limitation	Injection is cut off when engine speed rises above 6900 rev/min. and is reinstated below this figure. This is simply to afford some protection against over revving of the engine and the damage that may be caused
Battery voltage correction	This is a correction in addition to all other functions in order to compensate for changes in system voltage. The voltage is converted every TDC and the correction is then applied to all injection period calculations. On account of the time taken for full current to flow in the injector winding and the time taken for the current to cease, a variation exists depending on applied voltage

LEARNING TASKS

➡ Look back at the key words. Explain each one to a friend, and/or write out a short description to keep as evidence.

➡ Make a simple sketch to show the inputs and outputs of a full engine management system.

➡ Examine a real system or components, if available, to become familiar with the system operation.

➡ Write a short explanation about the claimed advantages of a full engine management system.

8 Diesel fuel injection

Introduction   The basic principle of the four-stroke diesel engine is very similar to the petrol system. The main difference is that only air is drawn into the engine's cylinders. The mixture formation takes place in the engine's combustion chamber as the fuel is injected under very high pressure. The timing and quantity of the fuel injected is important because of the usual issues of performance, economy and emissions. Fuel is metered into the combustion chamber by way of a high-pressure pump connected to injectors via heavy duty pipes. When the fuel is injected, it mixes with the air in the cylinder and will self-ignite at about 800°C.

The mixture formation in the cylinder is influenced by the following factors:

- Start of delivery and start of injection (timing). The timing of a diesel fuel injection pump to an engine is usually done using start of delivery as the reference mark (spill timing). The actual start of injection, in other words when fuel starts to leave the injector, is slightly later than start of delivery, as this is influenced by the compression ratio of the engine, the compressibility of the fuel and the length of the delivery pipes. This timing has a great effect on the production of carbon particles (soot), if too early and increases the hydrocarbon emissions, if too late.

- Spray duration and rate of discharge (fuel quantity). The duration of the injection is expressed in degrees of crankshaft rotation or in milliseconds. This clearly influences not only fuel *quantity*, but also the *rate* of discharge. This rate is not constant due to the mechanical characteristics of the injection pump.

Diesel engines do not in general use a throttle butterfly. Instead, the throttle acts directly on the injection pump to control fuel quantity. At low speeds in

Fuel-injection system with distributor-type pump
1 Fuel tank, 2 Fuel line, 3 Fuel filter,
4 Distributor-type fuel-injection pump,
5 Pressure line, 6 injection nozzle,
7 Fuel-return line.

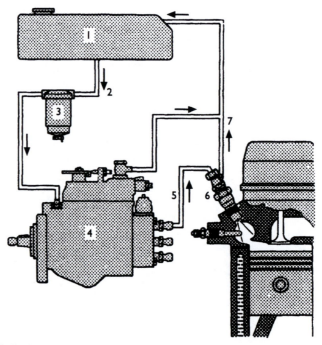

Figure 5.35 Fuel-injection system with distributor-type pump

particular, a very high excess air factor (very weak mixture) ensures complete burning and very low emissions. Where possible diesel engines operate with an excess air factor even at high speeds.

Pressure of injection will effect the quantity of fuel but the most important issue here is the effect on atomisation. At higher pressures, the fuel will atomise into smaller droplets with a corresponding improvement in the burn quality. Indirect injection systems use pressures up to about 350 bar and direct injection systems can be up to about 1000 bar. Emissions of soot are greatly reduced by higher pressure injection. Figure 5.35 shows a typical diesel fuel injection system.

Combustion   Diesel engines are called compression ignition engines. This is because the air in the cylinder is heated so quickly during the compression stoke that when fuel is injected, it self-ignites. Fuel is controlled or metered by the fuel-injection pump. It is injected into the combustion chamber through injection nozzles under very high pressure.

Fuel injection systems must have the following fuel delivery abilities:

- precise quantity according to engine load and speed
- correct instant relative to crankshaft position
- method appropriate to the combustion principle used.

On the basis of operating conditions such as engine speed and accelerator position, the system must determine the correct quantity of fuel to be injected. Fuel is injected into a pre-combustion chamber or directly into the main combustion chamber as shown in Figure 5.36.

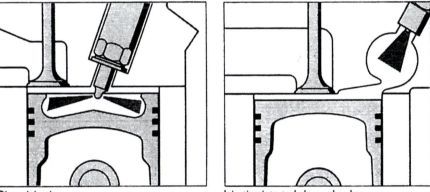

Direct injection   Injection into turbulence chamber

Figure 5.36 Diesel – direct and indirect injection

When hydrocarbon fuels are burned completely, carbon dioxide and water are the resulting products. Burning in internal-combustion engines is never quite complete. Varying load and speed conditions, poor mixing of air and fuel, and incorrect temperatures in the combustion chamber lead to partial burning. This causes other reactions that form toxic exhaust products.

Because diesel engines operate with excess air, CO emissions are very low. However HC's are a diesel problem under low-load and cold-running conditions. NOx forms due to high temperatures, which occur by way of pressure peaks during combustion. This is principally at high speeds and loads. The emission known as 'Diesel smoke' is a result of localised oxygen shortage. Splitting of pure hydrocarbons creates a cloud of tiny particles, called particulate emissions. This forms the familiar black smoke.

Basic diesel fuel injection   For each power stroke, the fuel-injection system must supply the required quantity of fuel to the injection nozzle at high pressure, and at precisely the right crankshaft position. The nozzle must spray finely atomised fuel into the combustion chamber. To ensure reduced emissions, the fuel must be metered so that there is excess air even at full load. The fuel quantity must also be controlled so that engine speed does not fall below the pre-set idle speed, nor exceed the maximum speed.

Injection timing is determined just as with petrol systems, according to engine speed and load. It is varied according to engine speed so that the main combustion pressure always occurs after the piston has passed TDC. Just enough fuel should be injected per degree of crankshaft rotation as can be burned in the same period of time.

The main part of the system is the fuel injection pump. This can be considered as having two main sections:

- The low-pressure section includes fuel tank, fuel filter, supply pump and overflow valve, as well as the various fuel lines.
- The high-pressure section, where the pressure necessary for injection is generated, is in the injection pump. Fuel is pumped through the pressure valve, high-pressure lines and nozzle holder to the injection nozzle.

For correct injection-pump operation, fuel must be continuously fed to the high-pressure section. The self-priming capacity of the supply pump, which is part of the distributor pump, may be adequate. However, an extra lift pump is often used.

The high-pressure section of the injection pump and the injection nozzles must be manufactured to tolerances of only a few thousandths of a millimetre. This means that even very tiny impurities in the fuel can interfere with operation. Poor filtration can, therefore, cause damage to the pump, pressure valves and injector nozzles. A fuel filter designed to a high standard is, therefore, necessary for long life and reliable operation.

Fuel sometimes contains water. If the water reaches the injection pump, it will damage it. For this reason, a fuel filter with a water trap from which the water must be drained periodically is used.

Diesel fuel injectors

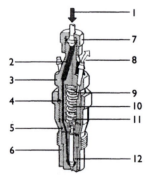

1. Fuel from injection pump
2. Filter
3. Valve-holder body
4. Pressure passage
5. Intermediate sleeve
6. Nozzle retaining nut
7. Union nut for pressure line
8. Leakage-fuel connection
9. Pressure adjusting shim
10. Pressure spring
11. Pressure spindle
12. Pintle nozzle

Figure 5.37 Nozzle holder with pintle nozzle

A nozzle holder secures the injector nozzle in the cylinder head and seals the combustion chamber. A high pressure injector pipe leads to the holder from the rotary fuel injection pump. A leak off fuel connection is also part of the nozzle holder. The holder assembly consists of the holder and injector nozzle.

The components of a nozzle holder include:

- body
- intermediate sleeve
- nozzle retaining nut
- pressure spindle
- spring
- adjusting shims.

The nozzle is centred in the body by the retaining nut. The body and retaining nut are screwed together to clamp the intermediate sleeve tightly against the body's flat face. The pressure spindle, spring and adjusting shim are inside the body. The spindle guides the spring and the nozzle needle centre the spindle. Inside the holder body, the high-pressure passage merges into the body's inlet bore, therefore connecting the injector with the injection pump via the high-pressure lines. A filter can also be installed in the pressure passage.

The pressure spring in the holder body acts on the nozzle needle via the pressure spindle. The spring preload determines the nozzle's opening pressure. This can be set by the adjusting shims. Fuel flows through holes in the holder body, sleeve and injection nozzle to the nozzle seat. During injection, fuel pressure lifts the needle and fuel flows through the orifice and the injection bores into the combustion chamber. When pressure falls off, the spring returns the needle to its seat, making a tight seal, and injection is completed.

The nozzle actually consists of two components, the nozzle body and needle. These two components are precisely machined to each other and must always be serviced as a pair.

In general, there are two injector types, but within these two main categories there are many detail variations appropriate to different engine configurations.

Multi-hole or orifice-type injectors for direct injection engines (Fig 5.38)	The orifice-type injector nozzle has a sealing cone on its body, a specially shaped nozzle seat and a 'blind hole'. These nozzles are usually of the multi-orifice type, but there are some single-orifice types. Single orifice types can have their injection orifice located centrally or on the side. Multiple orifice types have symmetrical or asymmetrical orifices. Nozzle opening pressures range between 150 and 250 bar

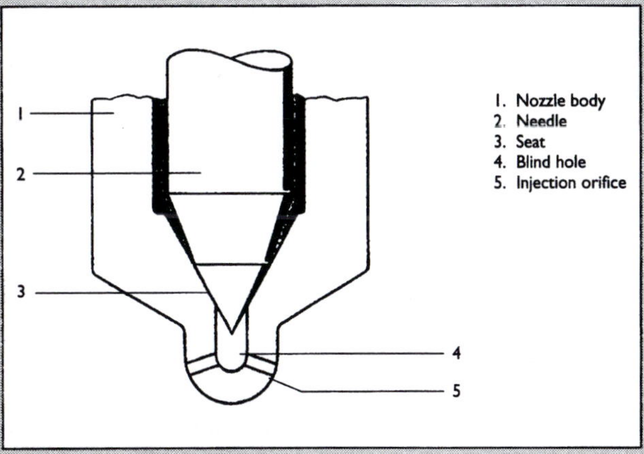

1. Nozzle body
2. Needle
3. Seat
4. Blind hole
5. Injection orifice

Figure 5.38 Cross section: orifice nozzle

Throttling pintle injectors for indirect injection engines (Fig 5.39)	Preparation of the fuel for combustion occurs mostly through air turbulence and the form of the injection spray. The nozzle opening pressure is often between 110 and 135 bar.

The needle has a specially-shaped pintle at one end for a pre-injection of fuel. As the needle first moves, only a very narrow annular orifice opens. This takes in just a little fuel. As the pressure forces it to open wider, the flow cross-section increases. The main quantity of fuel is injected only when the needle is almost fully open. Because it causes a gradual pressure rise, this injection method creates a soft combustion and, therefore, a relatively smooth engine operation. The spray pattern has a conical shape

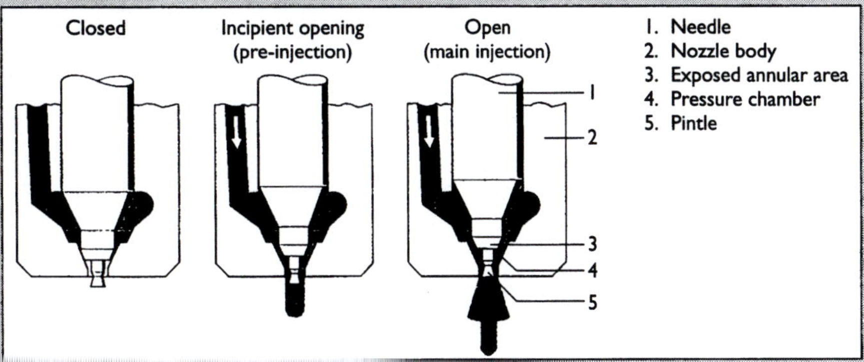

Closed   Incipient opening (pre-injection)   Open (main injection)

1. Needle
2. Nozzle body
3. Exposed annular area
4. Pressure chamber
5. Pintle

Figure 5.39 Throttling pintle nozzle: injection characteristics

Injection pressure lines, or injector pipes, are designed for the particular injection process and must not be altered during service. These lines connect the injection pump with the nozzle holders. The pipes are normally secured with clamps at predetermined intervals.

9 Bosch VE rotary injection pump – case study

Introduction

KEY WORDS

■ All words in the tables

■ Governor

■ Advance mechanism

■ Start of injection

■ Injection and ignition lag

■ Atomisation

■ Excess air factor

The Bosch VE distribution injection pumps meet the needs of a wide range of applications. A diesel engine's rated speed, power and configuration set the detail for a particular pump design. Most light vehicle diesel engines use a form of rotary pump, many now with electronic control. The VE pump is a well known and respected design and forms the basis of many electronic diesel control (EDC) systems.

The VE distribution pump has one pump cylinder and plunger regardless of the number of engine cylinders. Fuel delivered by a plunger is apportioned by way of a distributor groove to a number of ports. The number corresponds to the number of engine cylinders. The distributor pump contains the following groups:

■ high pressure pump with distributor

■ governor

■ timing mechanism

■ vane-type supply pump

■ shut off mechanism.

Figure 5.40 shows these groups and their functions.

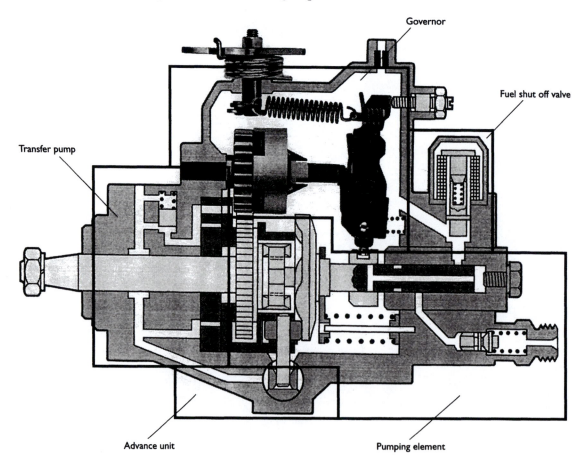

Figure 5.40 Rotary injection pump

The distributor fuel injection pump is engine-driven and runs at half crankshaft speed. It is positively driven so that the input shaft is synchronous with the engine. This is done by toothed belt drive when the engine is used for light vehicle applications.

Low-pressure fuel delivery

Bosch VE fuel injection systems have a vane-type fuel supply pump that draws fuel from the fuel tank and delivers it to the injection-pump. The supply pump delivers a virtually constant flow of fuel. A pressure-control valve is used to ensure a defined pressure level proportional to engine speed. Pressure can be set with this valve, to correspond to a particular engine speed. A small amount of fuel flows through the valve and returns to the suction side.

Pump-cavity pressure increases with rising engine speed

The supply pump surrounds the input shaft. Its impeller is concentric with the shaft and driven by a woodruff key. The impeller runs inside an eccentric ring fixed to the housing. The vanes are forced outward by centrifugal force against the eccentric ring. The fuel between the vanes' undersides and the impeller reinforces this.

The pressure-control valve is mounted close to the supply pump. It is a simple, spring-loaded slider valve that varies the pump-housing pressure according to the quantity of fuel being delivered. If fuel pressure exceeds a pre-set value, the valve piston opens the return circuit. This, in turn, allows fuel to flow back to the suction side of the pump. An overflow restriction screwed into the governor cover of the VE distributor pump allows a variable quantity of fuel to return to the fuel tank through a small bore.

High-pressure fuel delivery

High-pressure fuel delivery is the key to diesel fuel injection. It can be summarised as 'The right quantity of fuel should be delivered to the right injector at the right time – in a short time!'

Rotation of the input-shaft is transferred to the distributor plunger by dogs on the input shaft and cam plate meshing with a yoke. The roller ring, cam plate and the rollers convert the pure rotation of the input shaft into rotary-reciprocating motion.

The plunger is forced up towards *its* TDC position by the cam surface. Two return springs force it back down again towards *its* BDC. The springs also prevent the cam plate separating from the rollers under high acceleration. To ensure that the plunger is not displaced from its central position, the return springs are precisely matched to each other.

As well as driving the pump plunger, the cam plate also has an effect on fuel-injection pressure and duration. The important factors are cam stroke and lift speed. These factors must be individually set, depending on combustion chamber design. A special cam-plate surface is therefore developed for different engines. It is mounted on the end face of the cam plate.

The distributor head, plunger and control spool are lapped in so precisely to each other that they seal even at very high pressures. The complete pump assembly, therefore, must only be replaced as a complete unit.

Fuel metering

The delivery of fuel is a process, made up of several stroke phases. The plunger generates the pressure necessary for injection into the engine cylinders. Figure 5.41 illustrates the metering of fuel to an engine cylinder by the distributor-plunger motion phases. On a four-cylinder engine, the plunger rotates a quarter of a turn as it moves from bottom to top dead centre.

As the plunger moves from top to bottom dead centre, its rotary-reciprocating motion opens the inlet bore in the distributor head. This is by means of a metering slit in the plunger. Pressurised fuel in the pump cavity flows through this bore into the *high-pressure* chamber at the end of the distributor plunger. As motion reverses and the plunger moves towards its TDC, the plunger closes the inlet bore.

Continued motion opens a precisely defined outlet bore in the distributor head. Pressure build-up in the high-pressure chamber and the interior bore now opens the pressure valve. This finally is forced through the injector pipe to the injection nozzle.

The working stroke is completed as soon as the transverse cut-off bore of the plunger reaches the edge of the control spool. After this point, no more fuel is delivered to the injector and the pressure valve closes the line. Fuel returns through the cut-off bore to the pump cavity as the plunger completes its travel to TDC. This phase of plunger motion is called the 'residual stroke'. As the plunger returns, its transverse bore is closed just as the next control slit opens the fuel inlet bore. The high-pressure chamber is filled with fuel once again, and the cycle continues for the next cylinder.

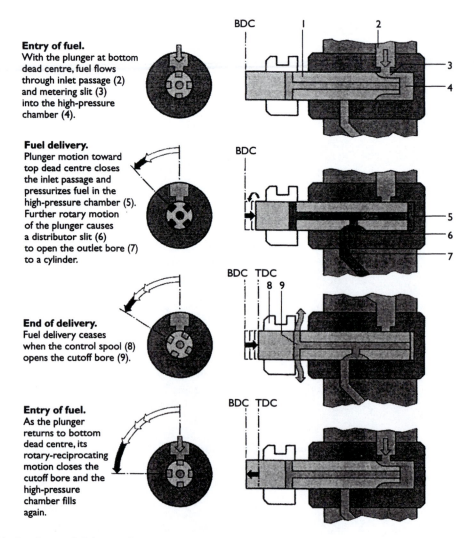

Entry of fuel.
With the plunger at bottom dead centre, fuel flows through inlet passage (2) and metering slit (3) into the high-pressure chamber (4).

Fuel delivery.
Plunger motion toward top dead centre closes the inlet passage and pressurizes fuel in the high-pressure chamber (5). Further rotary motion of the plunger causes a distributor slit (6) to open the outlet bore (7) to a cylinder.

End of delivery.
Fuel delivery ceases when the control spool (8) opens the cutoff bore (9).

Entry of fuel.
As the plunger returns to bottom dead centre, its rotary-reciprocating motion closes the cutoff bore and the high-pressure chamber fills again.

Figure 5.41 Strokes and delivery phases

A pressure valve (Figure 5.42) is used to close off the injection line from the pump. It also relieves the injection line pressure by removing a defined volume of fuel after the delivery phase. Therefore, it helps to determine the precise point at which the injector nozzle closes at the end of injection. The valve is a fluid-controlled plunger type. It is opened by fuel pressure and closed by a spring. Between delivery strokes, it remains closed. During delivery, the valve is lifted from its seat by the high fuel pressure. Fuel can then flow via slits into a ring groove and through the pressure-valve body, the pressure line and finally to the injection nozzle in the nozzle holder. As delivery reaches its end, pressure on the high-pressure side falls to that of the pump cavity and the pressure valve is closed by its spring.

a. closed
b. open
1. Valve holder
2. Valve seat
3. Valve spring

4. Valve body
5. Shaft
6. Relief piston
7. Ring groove
8. Longitudinal groove

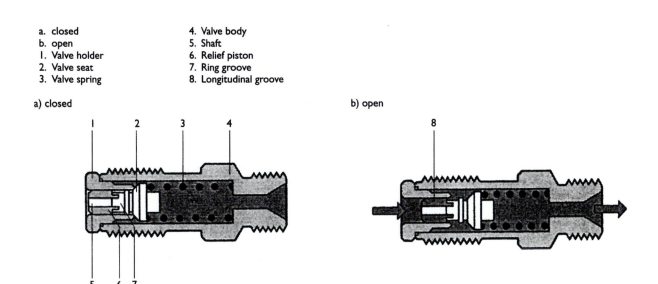

Figure 5.42 Pressure valve

Engine speed control

A diesel-powered vehicle can be said to have satisfactory driveability when its engine obeys the driver commands through the accelerator pedal.

- When starting off from rest, the engine should not tend to stall.
- When the pedal is moved, it must respond by accelerating or decelerating without bucking or lag.
- If the pedal is held steady on level ground or a constant slope, the vehicle speed should also remain steady.
- When the pedal is released, there should be engine braking.

It is the governor of the fuel-injection pump that tries to make sure all these conditions are met. The control strategy can be summarised in this way:

- Idle-speed regulation ensures that engine speed does not fall below the controlled idle speed.
- Maximum no-load speed control is such that, when load is removed, maximum full-load speed does not increase beyond the pre-set limit.
- In variable-speed governors, there is also intermediate speed control. Engine speeds between idle and maximum can be controlled within certain limits.

In addition to these, the governor can have other control functions for which add-on modules are required:

- releasing or blocking the extra quantity of fuel necessary for starting

■ altering full-load fuel delivery according to engine speed (torque control).

The ability of a governor to control engine speed is known as its 'speed droop'. This is the increase that occurs in engine speed (as a percentage) when loading is removed from a diesel engine, but with no change in accelerator position. In the control range, this speed increase should not exceed a set value. The controlled upper speed limit is the maximum value and occurs when the engine, starting at its maximum full-load speed, is relieved of all loading. The desirable speed droop characteristic depends upon engine application. For motor vehicles, a large speed-droop value is desired because it results in more stable speed control when small load changes take place, such as during acceleration or deceleration, and therefore gives better driveability.

The variable-speed governor controls all engine speeds between starting and maximum values. According to speed droop, the accelerator pedal can, therefore, set a constant engine speed.

Variable speed governor

Figure 5.43 shows the arrangement of a variable-speed governor's components. The input shaft drives the governor assembly.

The governor assembly consists of a housing, the centrifugal weights themselves as well as the governor spring and level assembly. This assembly is mounted in the housing so that it can rotate around the fixed governor axis. Radial motion of the centrifugal weights is translated into axial movement of a sliding sleeve.

The force acting on this sleeve and its resulting position change the position of the governor mechanism.

The governor mechanism consists of control lever, tensioning lever and starting lever; the control lever is pivoted in the pump housing and can be adjusted by the fuel-delivery adjusting screw. The starting and tensioning lever also rotate relative to the control lever.

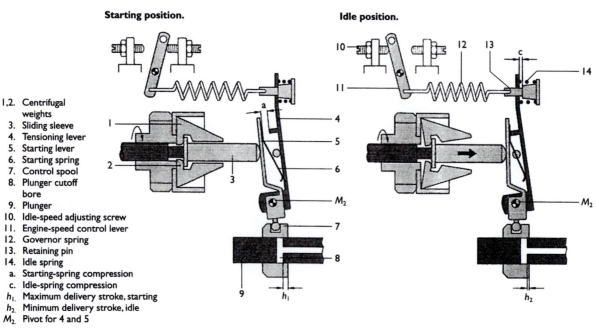

Starting position.   **Idle position.**

1,2. Centrifugal weights
3. Sliding sleeve
4. Tensioning lever
5. Starting lever
6. Starting spring
7. Control spool
8. Plunger cutoff bore
9. Plunger
10. Idle-speed adjusting screw
11. Engine-speed control lever
12. Governor spring
13. Retaining pin
14. Idle spring
a. Starting-spring compression
c. Idle-spring compression
h_1. Maximum delivery stroke, starting
h_2. Minimum delivery stroke, idle
M_2. Pivot for 4 and 5

Figure 5.43 Variable-speed governor

The starting lever has a ball pin on its underside, which engages with the control spool. The starting spring is attached to its upper end. At the upper end of the tensioning lever is a retaining pin over which is fitted the idle spring. The governor spring is also hooked into the end of the retaining pin. A lever and the control-lever shaft connect to the engine-speed control lever. The governor position is defined by the interaction of spring and sleeve force.

Control movement is transferred to the control spool, thus determining the quantity of fuel delivered by the distributor plunger. The variable speed governor can be considered to have a number of operating modes:

Starting	The engine is stationary and the centrifugal weights and sliding sleeve are in their rest position. The starting lever is moved to the starting position by its spring and pivots around the point. The control spool on the distributor plunger is held in its starting position by the ball pin of the starting lever. This set of conditions allows the plunger to have a large stroke. Maximum delivery is used for starting. Once cranking begins, the cranking speed starts to affect the centrifugal weights and moves the sliding sleeve against its weak spring. The starting lever again pivots, reducing delivery to the idle setting
Idle-speed	The engine-speed control lever is in its idle position, resting against the idle adjusting screw. Control is by the idle spring on the retaining pin, which balances the force from the centrifugal weights. This balance determines the sliding sleeve's position with reference to the plunger cut-off bore and, therefore, the length of the delivery stroke. At speeds above idle, the spring is collapsed and no longer has any effect
Under load	In operation, the engine-speed control lever assumes a position corresponding to a given engine speed. The driver controls this by pressing on the accelerator pedal. At speeds above idle, the starting and idle springs are collapsed and have no more effect. By pressing the accelerator pedal, the driver moves the engine speed control lever, thus stretching the governor spring by a certain amount. The result being that spring force is now greater than that of the centrifugal weights. Governor–spring tension pulls the starting tensioning levers. This moves the control spool into a position that will increase fuel delivery. The result is increased engine speed. The centrifugal weights now move out due to the increased speed, and move the sliding sleeve against the governor–spring tension. The control spool will remain in it *full* position until balance is regained. If engine speed continues to increase, the weights move further out and the sleeve force becomes stronger than the spring. The starting and tensioning levers now pivot the other way, moving the control spool in its *stop* direction so that the cut-off bore is opened earlier.
	This means that, for each position of the engine-speed control lever, there is a precise engine-speed range between full load and no load. The result is that the governor maintains the engine speed dictated by the accelerator pedal working within the parameter of its speed droop. If load increases to a point where the spool is in its full-load position but engine speed continues to fall, the weights move inward. However, because the spool is already at its maximum delivery position, fuel delivery cannot be increased further. The engine, is therefore, overloaded and the driver must change down a gear or ease off the accelerator
Engine-braking	When the vehicle drives the engine – such as when going down a hill or on over-run – the engine speed tends to increase. The governor weights move outward and the sleeve pushes against the starting and tensioning levers. The levers move the control spool toward less fuel delivery. This continues until the reduced delivery corresponds to the new load condition, at the extreme delivery falls to zero

Idle and maximum speed governor

The idle-maximum-speed governor controls only idle and maximum speeds. Between the two speeds, the accelerator pedal *directly* influences engine speed.

The governor assembly with its centrifugal weights and arrangement of levers is similar to that of the variable-speed governor previously described. The main difference is the governor spring and how it is installed. In this case the spring is a compression spring mounted inside a guide. A retaining pin connects the tensioning lever and governor spring. The idle and maximum speed governor can be considered to have a number of operating modes:

Starting	The engine is stationary and the centrifugal weights and sliding sleeve are in their rest position. The starting spring presses the starting lever against the sleeve. The control spool on the distributor plunger is, therefore, in the starting position
Idle control	When the engine has started and the accelerator pedal is released, the engine speed control lever is pulled back to idle position by the return spring. An increase in engine speed causes the centrifugal weights to move outward. This presses the sliding sleeve against the starting lever. The controlling element is the idle spring acting on the tensioning lever. As the starting lever pivots, the sleeve is moved in the *less delivery* direction. The balance between spring pressure and centrifugal force determines its position
Operation under load	When the driver presses the accelerator pedal, the engine-speed control level is pivoted by a certain amount. This collapses the starting and idle springs and the intermediate spring takes over. The latter spring allows a wide idle speed range, a large speed droop and a soft transition to the uncontrolled range. Further movement of the engine-speed control lever by the accelerator pedal collapses the intermediate spring so that the retaining-pin head now presses directly against the tensioning lever. The intermediate spring is now inactive and the mechanism is in its non-controlling range. The accelerator pedal now directly moves the sliding sleeve. To accelerate or climb a hill, the driver presses the pedal further, if less engine power is needed, he or she eases off. If engine load decreases with no change in pedal position, engine speed will increase without any increase in fuel delivery. The centrifugal force on the weights rises and the weights press the sliding sleeve against the starting and tensioning levers
Maximum speed	When the governor spring's tension is overcome by sleeve force, maximum-speed control sets in. This is designed to occur near the engine's rated speed. If the engine is relieved of all load, engine speed falls to its upper idle-speed limit and the engine is thus protected from over speeding

Injection timing

In order to compensate for injection and ignition lag, the timing-advance device can advance injection timing, relative to the crankshaft position, as the engine speed increases. The injection nozzle is opened by a pressure wave that moves through the high pressure injection pipe at the speed of sound. The time it takes for this to occur is not related to engine speed. However, as engine speed increases, the angle of crankshaft travel between the *start of pump delivery* and the *start of actual injection* at the nozzle also increases. Injection timing must therefore be advanced with increasing engine speed.

The start of injection from the injector lags behind the beginning of delivery from the pump. The size of the pipe and the speed of sound determine propagation time for the pressure wave. In diesel fuel, this is approximately 1500 m/s. The interval represented by the propagation time is referred to as *injection lag*. Because propagation time is virtually constant, the injection nozzle opens later at high engine speeds than at low engine speeds.

After the injection process, it takes a certain time for the fuel to atomise and mix with the air to form a combustible mixture (known as the delay period). This interval is independent of engine speed and is called *ignition lag*. Ignition lag is influenced by the fuel's combustibility, compression ratio, air temperature and fuel atomisation. It only amounts to about one millisecond in most cases.

If the beginning of injection is not advanced as engine speed increases, the crankshaft rotates through a larger and larger angle between the injection point and the beginning of combustion. This would mean that combustion no longer starts at the right instant as far as piston position is concerned. Because a diesel engine produces the best power output when combustion is correctly timed, injection timing must be advanced with increasing engine speed to compensate for injection and ignition lag.

The hydraulic timing-advance device (Figure 5.44) is built into the distributor pump's underside. On one side of the piston is a fuel inlet bore, on the other side a spring. A sliding block and actuating pin connect the piston with the roller ring. The timing piston is held in its initial position by a preloaded spring. During operation, fuel pressure in the pump housing is regulated proportional to engine speed by the pressure-control valve. As a result, the piston side opposite to the spring is under the same proportional pressure.

When engine speed reaches about 300 rev/min., fuel pressure reaches a value sufficient to overcome the spring preload and move the timing piston. This piston motion is transmitted via the sliding block and pin to the roller ring, which rotates in bearings. This in turn alters the arrangement of cam plate and roller ring so that the rotating cam plate is lifted at an earlier point in time by the rollers. The rollers with their ring are thus turned by a specific angle relative to cam plate and distributor plunger. This angle can be as much as 12° of camshaft rotation or 24° of crankshaft rotation.

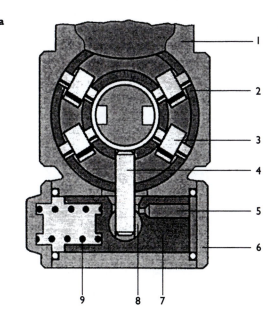
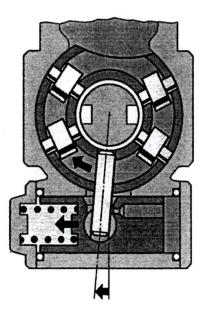

a. Position with engine at rest
b. Operational position
1. Pump housing
2. Roller ring
3. Rollers
4. Actuation pin
5. Bore in timing piston
6. Cover
7. Timing piston
8. Sliding block
9. Spring

Figure 5.44 Operation of timing-advance device

Electronic control of diesel injection

The advent of electronic control over the diesel injection pump has allowed many advances over the purely mechanical system. The production of high pressure and injection is, however, still mechanical with all current systems. The following advantages are apparent for the electronic, when compared to the non-electronic, control system.

- more precise control of fuel quantity injected
- better control of start of injection
- better idle speed control
- control of exhaust gas recirculation
- drive by wire system (potentiometer on throttle pedal)
- anti-surge function
- output to data acquisition systems, etc.
- temperature compensation
- cruise control.

Because fuel must be injected at high pressure, the hydraulic head, pressure pump and drive elements are still used. An electromagnetic moving iron actuator adjusts the position of the control collar, which, in turn, controls the delivery stroke and therefore the injected quantity of fuel. Fuel pressure is applied to a roller ring, and this controls the start of injection. A solenoid-operated valve controls the supply to the roller ring. These actuators together allow control of the start of injection and injection quantity.

Ideal values for fuel quantity and timing are stored in memory maps in the electronic control unit. The injected fuel quantity is calculated from the accelerator position and the engine speed. Start of injection is determined from fuel quantity, engine speed, engine temperature and air pressure. The ECU is able to compare start of injection with actual delivery from a signal produced by the needle motion sensor in the injector.

Control of exhaust gas recirculation can be undertaken by using a simple solenoid valve. This is controlled as a function of engine speed, temperature and injected quantity. The ECU is also in control of the stop solenoid and glow plugs via a suitable relay.

Figure 5.45 shows a typical layout of an electronic diesel control system.

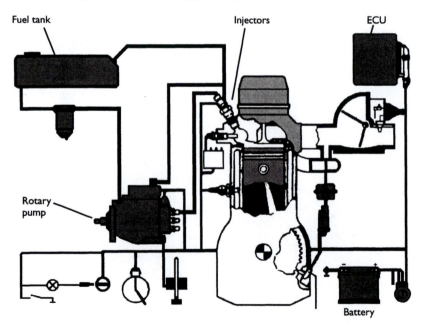

Figure 5.45 Electronic control of diesel injection with distributor pump

10 Other engine management functions

ECU self-diagnostics

Most ECU's are now equipped to advise the driver of a fault in the system, and to aid the repairer in detection of the problem. The detected fault is first notified to the driver by a dashboard warning light. A code giving further details is held in RAM within the ECU. As an aid to fault finding, the repairer can read this fault code using suitable equipment.

Each fault detected is memorised as a numerical code and can only be erased by the repairer. If the fault is not detected again for fifty starts of the engine, the ECU often erases the code automatically. Only serious faults will light the warning lamp but minor faults are still recorded in memory. The faults are memorised in the order they occur.

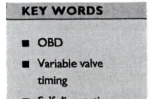

KEY WORDS

■ OBD

■ Variable valve timing

■ Self diagnostics

■ Fault code or blink code

Certain major faults will cause the ECU to switch over to an emergency mode. In this mode, the ECU substitutes alternative values in place of the faulty signal. This is designed as a 'limp home' facility.

Faults can be read as two digit numbers from the flashing warning light by earthing an appropriate wire in the diagnostic link for between 2.5 seconds and 10 seconds. Earthing this wire for more than ten seconds will erase the fault memory, as does removing the ECU constant battery supply. Note that earthing of a wire to read fault codes should only be carried out in accordance with the manufacturers' recommendations. The same coded signals can be read on many after sales service testers.

Figure 5.33 on page 90 shows the Bosch Motronic M5 with the on board diagnostic or OBD II system. On board diagnostics are becoming essential for the long-term operation of a system's ability to produce a clean exhaust. In the USA, a very comprehensive diagnosis of all components in the system which affect the exhaust is now required. It can be expected that a similar requirement will be made within the EU. Any fault detected must be indicated to the driver by a warning light.

Digital electronics allow both sensors and actuators to be monitored. Allocating a range of operating values to all of the sensors and actuators does this. If a deviation from these figures is detected, this is stored in memory and can be output in the workshop to assist with fault finding. A warning light is also illuminated.

Monitoring of the ignition system is very important as misfiring not only produces more emissions of hydrocarbons but the unburnt fuel enters the catalytic converter and burns there. This can cause higher than normal temperatures and may damage the catalytic converter.

An accurate crankshaft speed sensor is used to monitor ignition and combustion in the cylinders. Misfiring alters the torque of the crankshaft for an instant, which causes irregular rotation. This allows a misfire to be recognised instantly.

A number of further sensors are required for the OBD II functions. Another lambda sensor *after* the catalytic converter monitors its operation. An intake pressure sensor and a valve are needed to control the activated charcoal filter to reduce and monitor evaporative emissions from the fuel tank. A differential pressure sensor also monitors the fuel tank permeability. Beside the driver's fault lamp, a considerable increase in the electronics is required in the control unit in order to operate this system.

Other areas of engine control

As the requirements for lower and lower emissions continues, together with the need for better performance, other areas of engine control are constantly being investigated. This is becoming even more important as the possibility of carbon dioxide emissions being included in the regulations increases. Some of the current and potential areas for further control of engine operation are included in this section.

Variable inlet tract

It is not possible for an engine to operate at its optimum volumetric efficiency with fixed manifolds. This is because the length of the inlet tract determines the velocity of the intake air and, in particular, the propagation of the pressure waves set up by the pumping action of the cylinders. These standing waves can be used to improve the ram effect of the charge as it enters the cylinder but only if they coincide with the opening of the inlet valves. The length of the inlet tract has an effect on the frequency of these waves. One method of changing the length of the inlet tract is shown in Figure 5.46.

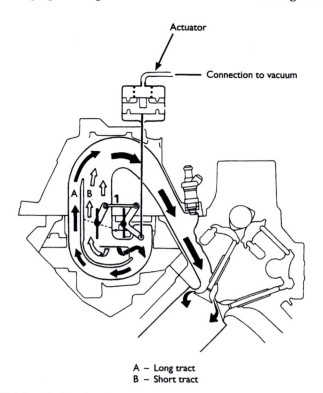

A – Long tract
B – Short tract

Figure 5.46 Variable length inlet tract

Variable valve control

With the more widespread use of twin cam engines, one cam for the inlet valves and one for the exhaust valves, it is possible to vary the valve overlap while the engine is running. Honda uses this technique, but also has a system which noticeably improves the power and torque range by opening one valve at low speed and two at high speed. A system of valves using oil pressure locks the valve to the cam. The ideal operation of the valves is determined from a suitable map held in ROM in the control unit. Figure 5.47 shows this system.

At low revs the VTEC-E engine opens only one inlet valve per cylinder fully, so just 12 valves control the mixture and combustion of air and fuel. This delivers maximum efficiency with lowest possible emissions. At higher engine speeds hydraulic pins activate the extra valves to give 16 valve performance

Figure 5.47 Honda's valve control system

Combustion and pressure sensing

Research is on-going in the development of cost effective sensors for determining combustion pressure and combustion flame quality. These sensors are often used during development and are now starting to be used in production vehicles. These sensors provide instant closed loop feedback about the combustion process. This is particularly important with lean burn engines.

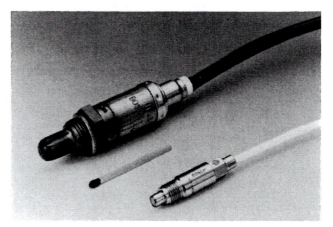

Figure 5.48 Lambda sensor and combustion pressure sensor

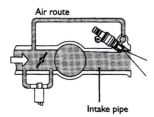

Figure 5.49 Injection valve with air shrouding

Injectors with air shrouding

If high-speed air is introduced at the tip of an injector, the dispersal of the fuel is considerably improved. Droplet size can be reduced to below 50 microns during idle conditions. Figure 5.49 shows an injector with air shrouding and Figure 5.50 shows a comparison of shrouded and non-shrouded injectors in operation.

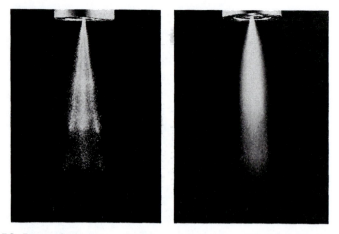

Figure 5.50 Better fuel preparation through injection with air shrouding. Left: Injection valve without air shrouding. Right: Injection valve with air shrouding

Simulation programs   A computer program that represents how the ECU calculates fuelling requirements and timing is a good way to learn the system operation. A Windows 95/98® shareware program is available from my web site. The address is at the front of the book.

LEARNING TASKS

➥ Look back at the key words. Explain each one to a friend, and/or write out a short description to keep as evidence.

➥ Make a simple sketch of a valve timing diagram to show the effect of changing the camshaft phasing.

➥ Examine other sources of information and find out how fault codes can be accessed.

➥ Write a short explanation about how OBD can help the repairer but may also cause further problems.

---

11   Emission control

Engine design   Many design details of an engine have a marked effect on exhaust emissions. With this in mind, the final engine design will be a compromise between conflicting interests. Some major areas of interest are listed below.

KEY WORDS

■ All words in the tables

■ Lambda

■ Lambda window

■ Catalytic converter

Combustion chamber design The major source of hydrocarbon emissions is from unburnt fuel, which condenses as it contacts the combustion chamber walls. For this reason, the surface area of the walls should be kept as small as possible and with the least complicated shape. The theoretical ideal is a sphere, but this is far from practical. Good swirl of the cylinder charge is important as this allows better and more rapid burning. Almost as important is to ensure a good swirl in the area of the spark plug. This ensures a quality of mixture that is easier to ignite. The spark plug is best positioned in the centre of the combustion chamber as this reduces the likelihood of combustion knock by reducing the distance the flame front has to travel

Compression ratio	In general, the higher the compression ratio, the higher the thermal efficiency of the engine. The two main drawbacks to higher compression ratios are the increased emissions and the increased tendency to knock. The problem with emissions is due to the higher temperature, which causes a greater production of NOx. An increase in temperature also makes the cylinder charge more likely to self-ignite. This results in combustion knock. Countries who have had stringent emission regulations for some time, such as the USA and Japan, have tended to develop lower compression engines. However, with changes in combustion chamber design, introduction of four valves per cylinder and greater electronic control, compression ratios have tended to increase over the years. Currently, the average value is approximately 9:1
Valve timing	The effect of valve timing on exhaust emissions can be quite considerable. One of the main factors being the amount of *valve overlap*. The inlet and exhaust valves are open together (at about TDC) during this time. This is the beginning of the induction stroke and the end of the exhaust stroke. The duration of this phase determines the amount of exhaust gas left in the cylinder when the exhaust valve finally closes. This internally-recirculated exhaust gas has a significant effect on the reaction temperature. The more exhaust gas left in the cylinder, the lower the temperature. This has a particular effect on the emissions of NOx. A down side is that a greater valve overlap at idle can greatly increase emissions of HC. This compromise problem has led to the introduction of electronically controlled valve timing
Manifold designs	Gas flow within the inlet and exhaust manifold is a very complex subject. The main cause of this complexity is the changes in flow, due not only to changes in engine speed but also to the suction and pumping action of the pistons in the cylinders. This pumping and suction action causes pressure fluctuations in the manifolds. If the manifolds and both induction and exhaust systems are designed to reflect the pressure wave back at just the right time, great improvements in volumetric efficiency can be attained. Many vehicles are now fitted with adjustable length induction tracts. Longer tracts are used at lower engine speeds and shorter tracts at higher speed
Charge stratification	If the charge mixture can be inducted into the cylinder in such a way that a richer mixture is in the proximity of the spark plug, then overall the cylinder charge can be much weaker. The idea is that the mixture near the plug will light and will then, in turn, be able to light a progressively weaker mixture across the chamber. This can bring great advantages in fuel consumption, but the production of NOx can still be a problem
Warm up time	A significant quantity of toxic emissions is created during the warm up phase. Suitable materials and care in the design of the cooling system can reduce this problem

Exhaust gas recirculation This technique is used to reduce combustion temperatures which, in turn, reduces the production of nitrogen oxides (NOx). EGR can be either internal as mentioned earlier due to valve overlap, or external via a simple arrangement of pipes and a valve (Figure 5.51). A proportion of exhaust gas is

Schematic diagram of exhaust-gas recirculation.

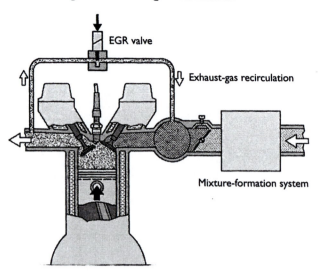

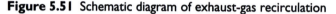

Figure 5.51 Schematic diagram of exhaust-gas recirculation

simply returned to the inlet side of the engine. This is controlled electronically as determined by the ROM in the ECU. This ensures that driveability is not effected and also that the rate of EGR is controlled. If the EGR rate is too large, the production of hydrocarbons increases. The main drawback of EGR systems is that they can become restricted by exhaust residue over a period of time, thus changing the actual percentage of recirculation.

Ignition system

The ignition system can effect exhaust emissions due to the:

- quality of the spark
- timing of the spark.

The quality of a spark will determine its ability to ignite the mixture. The duration of the spark in particular is significant when igniting weaker mixtures. A stronger spark is less likely to cause a misfire. A misfire will cause massive increases in the production of hydrocarbons.

The timing of a spark is clearly critical but, as ever, it is a compromise between power, driveability, consumption and emissions. Figure 5.3 on page 66 shows the influence of ignition timing on emissions and fuel consumption. The production of carbon monoxide is dependent on fuel mixture and is not significantly effected by changes in ignition timing. Electronic and programmed ignition systems have made significant improvements in the reduction of emission levels of today's engines.

Thermal after-burning

Prior to the more widespread use of catalytic converters, thermal after-burning was used to reduce the production of hydrocarbons. In fact, hydrocarbons do continue to burn in the exhaust manifold, and recent research has shown that the type of manifold used (such as cast iron or pressed steel) can have a noticeable effect on the reduction of HC. At temperatures of about 600°C, HC and CO are burnt or oxidised into H_2O and CO_2. If air is injected into the exhaust manifold just after the valves, then the after-burning process can be encouraged.

Catalytic converters

Further, more stringent, regulations in most parts of the world have made the use of a catalytic converter almost inevitable. The three-way catalyst (TWC) is used to great effect by most manufacturers. It is a very simple

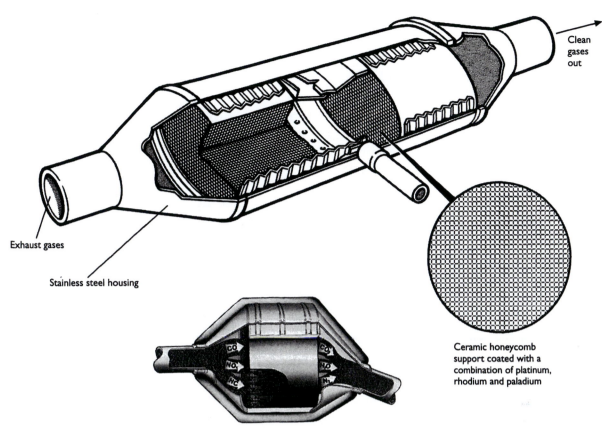

Clean gases out

Exhaust gases

Stainless steel housing

Ceramic honeycomb support coated with a combination of platinum, rhodium and paladium

Basic chemistry of catalytic converter

Figure 5.52 Catalytic converter

device, looking similar to a standard exhaust box. Note that, in order to operate correctly, the engine must be run at or very near to stoichiometry (the ideal air/fuel ratio of 14.7:1). This is to ensure that the right 'ingredients' are available for the catalyst to perform its function. Figure 5.52 shows a magnified view of the inside of a catalytic converter. The balanced chemical equations encouraged by the catalyst are as follows:

- $2CO + O_2 = 2CO_2$
- $2C_2H_6 + 7O_2 = 4CO_2 + 6H_2O$
- $2NO + 2CO = N_2 + 2CO_2$

There are, in fact, many types of hydrocarbons, but the above example illustrates the reaction. Note that the reactions rely on some CO being produced by the engine in order to reduce the NOx. This is one of the reasons that manufacturers have been forced to run engines at stoichiometry (the ideal air fuel ratio). This legislation has tended to stifle the development of lean burn techniques. The fine details of the emission regulations can, in fact, have a very marked effect on the type of reduction techniques used.

Figure 5.52 shows a ceramic monolith type of base for the catalyst material. This ceramic is a magnesium aluminium silicate and, due to the several thousand very small channels, provides a large surface area. This area is coated with a wash coat of aluminium oxide, which increases its effective surface area by about seven thousand. Noble metals are used for the catalysts. Platinum promotes the oxidation of HC and CO, and rhodium helps the reduction of NOx. The whole three-way catalytic converter contains about three to four grams of the precious metals.

The ideal operating temperature range is from 400°C to 800°C, and the delay in the catalyst reaching this temperature is a problem. This is known as 'catalyst light off time'. Various methods have been used to reduce this time, because a lot of emissions are produced before light off occurs. Electrical heating is one solution as is a form of burner, which involves lighting fuel inside the converter. A small electrically heated pre-converter is shown as Figure 5.53. Initial tests for this system show that the emissions of hydrocarbons during the warm up phase can be reduced significantly. The problem is that about 30 kW of heat are required during the first 30 seconds to warm up the pre-converter. This will require a current in the region of 250 A.

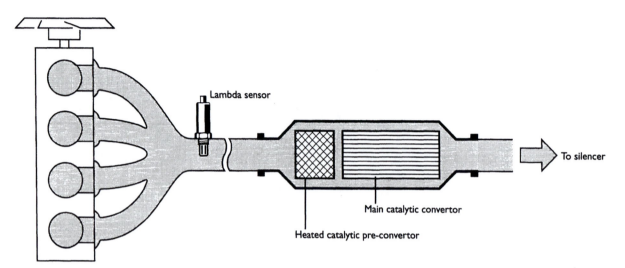

Figure 5.53 Electrically heated catalytic pre-converter

Catalytic converters can be damaged by the engine in two ways:

- by the use of leaded fuel which causes lead compounds to be deposited on the active surfaces, thus reducing effective area
- by engine misfire which can cause the catalytic converter to overheat due to burning inside the unit.

Some manufacturers, BMW for example, use a system on some vehicles where a sensor monitors output of the ignition HT system and, if the spark is not present, will not allow fuel to be injected.

Diesel exhaust emissions

Overall, the emissions from diesel combustion are far lower than emissions from petrol combustion. Figure 5.54 shows the comparison between petrol and diesel emissions. The CO, HC and NOx emissions are lower mainly due

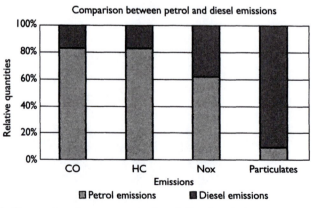

Figure 5.54 Comparison between petrol and diesel emissions

to the higher compression ratio and excess air factor. The higher compression ratio improves the thermal efficiency and thus lowers the fuel consumption. The excess air factor ensures more complete burning of the fuel.

The main problem area is that of particulate emissions. These particle chains of carbon molecules can also contain hydrocarbons. The dirt effect of this emission is a pollution problem, but the possible carcinogenic effects of this soot also gives cause for concern. The diameter of these particles is only a few ten thousandths of a millimetre. This means they float in the air and can be inhaled.

Exhaust emissions from diesel engines have been reduced considerably by changes in the design of combustion chambers. The more accurate control of start of injection and spill timing has also allowed improvements to be made. Electronic control, as discussed in a previous section, has made a significant contribution. A number of further techniques can be employed to control emissions. These are set out in the following table.

Exhaust gas recirculation	In much the same way as with petrol engines, EGR is employed primarily to reduce NOx emissions by reducing the reaction temperature in the combustion chamber. However, if the percentage of EGR is too high, increased hydrocarbons and soot are produced
Intake air temperature	This is appropriate to turbocharged engines such that if the air is passed through an intercooler as well as improvements in volumetric efficiency, the lower temperature will again reduce the production of NOx
Catalytic converter	A catalyst can be used to reduce the emission of hydrocarbons. The reduction of nitrogen oxides by a catalyst is also possible due to recent developments. Using a catalytic converter with a diesel is difficult because of the low CO production
Filters	To reduce the emission of particulate matter (soot), filters can be used. These can vary from a fine grid design made from a ceramic material to centrifugal filters and water trap techniques

Closed loop lambda control   Current EU regulations have all but made closed loop control of air fuel mixture in conjunction with a three way catalytic converter mandatory.

Lambda control is a simple closed loop feedback system in that the signal from a lambda sensor in the exhaust can directly affect the fuel quantity injected. Figure 5.55 shows a block diagram of the lambda control system.

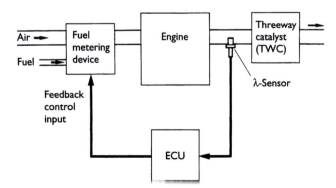

Figure 5.55 Fuel metering with closed loop control

The principle of operation is as follows: the lambda sensor produces a voltage which is proportional to the oxygen content of the exhaust which is, in turn, proportional to the air–fuel ratio. At the ideal setting, this voltage is about 450 mV. If the voltage received by the ECU is below this value, indicating a weak mixture, the quantity of fuel injected is increased slightly. If the signal voltage is above this value, indicating a rich mixture, the fuel quantity is reduced.

A delay also exists between the mixture formation in the manifold and the measurement of the exhaust gas oxygen. This is due to the engine's working cycle and the speed of the inlet mixture, the time for the exhaust to reach the sensor and the sensors response time. This is sometimes known as 'dead time' and can be as much as one second at idle speed but only a few hundred milliseconds at higher engine speeds.

Due to the dead time the mixture cannot be controlled to an exact value of lambda equals one. It is, however, possible to keep the mixture in the lambda window of 0.97 to 1.03 which is the region in which the catalytic converter is at its most efficient.

Other sources of emissions

Besides the exhaust, there are two other main sources of emissions on a vehicle:

- crankcase emissions from the engine
- evaporative emission from the fuel tank and system.

Crankcase emissions are due to combustion gases being forced past the piston. These are dealt with by using positive crankcase ventilation. Refer to the level two book for more information on this subject.

Evaporative emissions, as the name suggests, are due to fuel evaporating. The main source of this is the fuel tank, particularly when the ambient temperature is high. The most common way to reduce this problem is to have a system of valves that keep the pressure in the fuel tank just below atmospheric pressure. Further, a charcoal canister is often used to trap or store the emissions. At a suitable time, a valve operated by the ECU purges the canister into the inlet manifold where the gases are subsequently burned.

LEARNING TASKS

- Look back at the key words. Explain each one to a friend, and/or write out a short description to keep as evidence.
- Examine a real system and note any components associated with reducing emissions.
- Write a short explanation about why legislation influences the design of engines and associated systems.

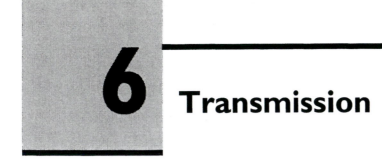

6 Transmission

1 Introduction

Start here! Do you remember the basic reason a transmission system is required? The power of an engine consists of speed and torque. Torque is the twisting force of the engine's crankshaft, and speed refers to its rate of rotation. The transmission can adjust the proportions of torque and speed that is delivered from the engine to the drive shafts. When torque is increased speed decreases, and when speed is increased the torque decreases. The transmission also reverses the drive and provides a neutral position when required. The layout of the transmission components varies depending on the drive configuration, FWD and RWD being the most common (Figures 6.1 and 6.2).

Figure 6.1 Front wheel drive layout

Figure 6.2 Rear wheel drive layout

KEY WORDS

■ All words in the
  table

The term 'transmission' refers to a series of parts that transmit power from the engine to the vehicle's drive wheels. The gearbox, drive shaft, and final drive make up the drive train of the vehicle. There are two main kinds of transmissions: manual and automatic. However, these two types come in a number of forms. Most transmissions vary torque and speed by means of gears but there are now systems known as 'CVT' or constantly variable transmission.

Terminology

Transmission	A collection of components to vary the speed and torque available at the drive wheels to allow operation of the vehicle under all normal conditions
Stall test	This test is to determine the correct operation of an automatic transmission torque converter and to ensure that there is no slip
4WD	A vehicle with four-wheel drive delivers power to all four wheels. A transfer box distributes the power between the front and rear wheels
CVT	Constantly variable transmission is a system where variable ratio pulleys are used instead of gears. The pulleys are connected by a special drive chain
Torque converter	This is a device that most automatic transmission systems have instead of a mechanical clutch. It delivers power from the engine to the transmission and also increases the torque when the car begins to move
Ratio	When one gear drives another, the speed and torque of the *driven* gear will vary depending on the ratio of the teeth on each gear. A gear with 20 teeth driving a gear with 40 teeth will produce an increase in torque of 1:2 (20:40) and a reduction in speed of 2:1 (40:20)
Lock up	Even when operating at their maximum efficiency, a torque converter will still slip just a little. A mechanical or hydraulic lock is used to ensure no slip occurs
Epicyclic gears	A set of gears used in many automatic transmission systems. They consist of a sun wheel, planet gears and an annulus
Torque	A turning force measured in Nm
Speed	Rate of rotation when referring to transmission normally measured in rev/min.
Power	The rate at which work is being done. In a transmission system, the power can remain the same but have different speed and torque. Low speed high torque and high speed low torque can be at the same power level
Detent	A mechanism in a gearbox to hold a gear in mesh with another
Synchromesh	Meshing gears or dog clutches is only possible when they are running at the same speed. Synchromesh synchronises the speeds before gear clutches are meshed
Interlock mechanism	A mechanism in a gearbox to only ever allow one gear to be in mesh at any one time

LEARNING TASKS

➡ Look back at the key words. Explain each one to a friend, and/or write out a short description to keep as evidence.

➡ Make a simple sketch to show a 20-tooth gear running at 1000 rev/min., driving a 15-tooth gear. Mark on the diagram the direction of rotation, the speeds and torques.

➡ Examine a real system and note the layout of transmission components.

2 Manual transmission

Clutch A clutch is a device for disconnecting and connecting rotating shafts. In a vehicle with a manual gearbox, the driver pushes down the clutch when changing gear to disconnect the engine from the gearbox. It also allows a temporary neutral position when, for example, waiting at traffic lights and a way of gradually taking up drive from rest.

The clutch is made of two main parts, a pressure plate and a driven plate. The driven plate, often termed the clutch disc, is fitted on the shaft, which takes the drive into the gearbox. When the clutch is engaged, the pressure plate presses the driven plate against the engine flywheel. This allows drive to be passed to the gearbox. Pushing down the clutch springs the pressure plate away, which frees the driven plate. Figure 6.3 shows a typical diaphragm-type clutch assembly. Earlier clutches and some heavy-duty types use coil springs instead of a diaphragm. Figure 6.4 shows how clamping force of a diaphragm type clutch varies as the plate wears and the diaphragm moves. Figure 6.5 shows the movement of the diaphragm during clutch operation. The diaphragm-type clutch replaced an earlier type with coil springs and it has a number of advantages when used on light vehicles:

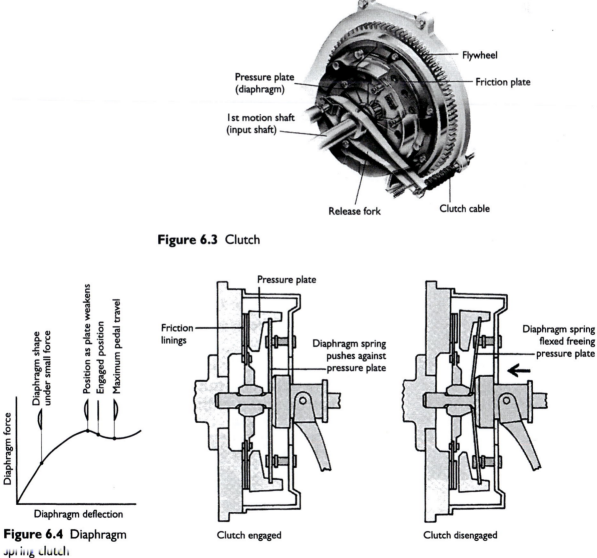

Figure 6.3 Clutch

Figure 6.4 Diaphragm spring clutch characteristics

Figure 6.5 Diaphragm clutch operation

- it is not affected by high speeds (coil springs can be thrown outwards)
- its low pedal force makes for easy operation
- it is light and compact
- clamping force increases, or at least remains constant, as the friction lining on the plate wears.

The method of controlling the clutch is quite simple. The mechanism consists of either a cable or hydraulic system as shown in Figure 6.6.

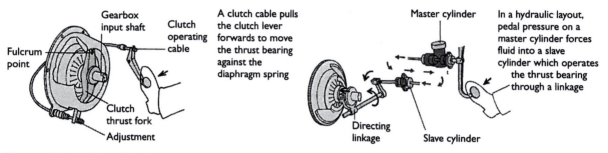

Figure 6.6 Cable and hydraulic clutch operation

Multiplate clutches   Multiplate clutches are used in specialist applications, such as for very high performance vehicles. Some motorcycles and heavy commercial vehicles also use clutches of this type. The principle is the same as a single plate clutch except that, with multiple plates, greater power can be transmitted. A common use, however, of a multiplate clutch is in an automatic gearbox. This is because a number of clutches are needed to control the gears. As space is limited, multiple plates are used to allow all the power to be transmitted. Modern limited slip differentials also make use of the multiplate clutch. Figure 6.17 shows a typical multiplate wet clutch. This example is as used in one type of automatic gearbox.

Manual gearbox   The driver changes the gears of a manual gearbox by moving a hand-operated lever called a gear stick. All manual gearboxes have a neutral position, three, four, or five forward gears, and a reverse gear. A few even have six forward gears now! The driver puts the gearbox into neutral as the engine is being started, or when a car is parked with the engine left running.

Power travels into the gearbox via the input shaft. A gear at the end of this shaft drives a gear on another shaft called the countershaft or layshaft. A number of gears of various sizes are mounted on the layshaft. These gears drive other gears on a third motion shaft also known as the output shaft.

The gearbox produces various gear ratios by engaging different combinations of gears. For reverse, an extra gear called an idler operates between the countershaft and the output shaft. It turns the output shaft in the opposite direction to the input shaft. Figure 6.7 shows a front wheel drive gearbox and Figure 6.8 shows the power flows through this box in each of the different gears. Note how in each case (with the exception of reverse) the gears do not move. This is why this type of gearbox has become known as 'constant mesh'. In other words, the gears are running in mesh with each other at all times. Dog clutches are used to select which gears will be locked to the output shaft. These clutches which are moved by selector levers, incorporate synchromesh mechanisms as shown in Figure 6.9 as part of a gearbox's input shaft components.

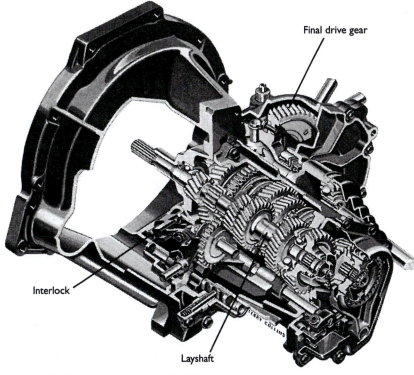

Figure 6.7 Five speed gearbox (FWD)

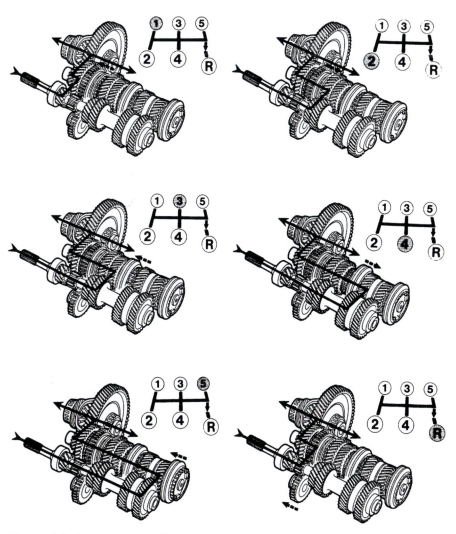

Figure 6.8 Gearbox power flows

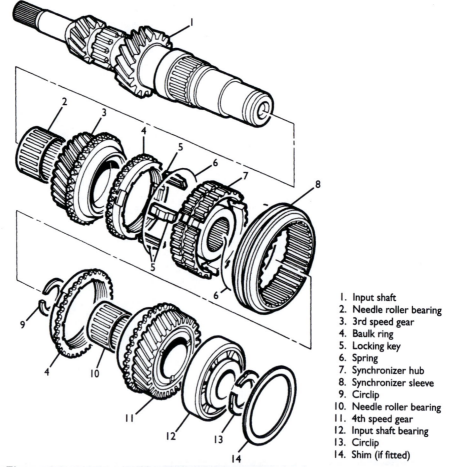

1. Input shaft
2. Needle roller bearing
3. 3rd speed gear
4. Baulk ring
5. Locking key
6. Spring
7. Synchronizer hub
8. Synchronizer sleeve
9. Circlip
10. Needle roller bearing
11. 4th speed gear
12. Input shaft bearing
13. Circlip
14. Shim (if fitted)

Figure 6.9 Gearbox input shaft components showing the synchronizer mechanism

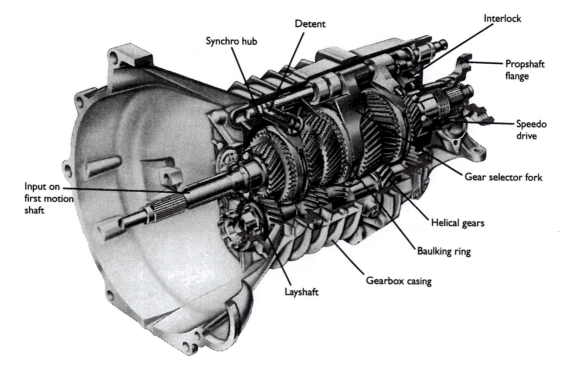

Figure 6.10 Rear wheel drive gearbox

A synchromesh mechanism is needed because the teeth of the dog clutches would clash if they met at different speeds. The system works like a friction type cone clutch. The collar is in two parts and contains an outer toothed ring that is spring loaded to sit centrally on the synchromesh hub. When the outer ring (synchronizer sleeve) is made to move by the action of the selector mechanism, the cone clutch is also moved because of the locking keys. The gear speeds up as the cones touch, thus allowing the dog clutches to engage smoothly. A baulking ring is fitted between the cone on the gear wheel and the synchronizer hub. This is to prevent engagement until the speeds are synchronized.

A detent mechanism is used to hold the selected gear in mesh. In most cases this is just a simple ball and spring acting on the selector shaft(s). Figure 6.10 shows a rear wheel drive gearbox with the detent mechanism marked. Gear selection interlocks are a vital part of a gearbox. These are to prevent more than one gear from being engaged at any one time. On the single rail (one rod to change the gears) gearbox shown in Figure 6.10, the interlock mechanism is shown at the rear. As the rod is turned (side to side movement of the gear stick) towards first-second, third-fourth or fifth gear positions, the interlock will only engage with either the first-second, third-fourth or fifth gear selectors as appropriate. Equally, when any selector clutch is in mesh, the interlock will not allow the remaining selectors to change position. Figure 6.11 shows a front wheel drive selector mechanism where two rails are used to select the gears.

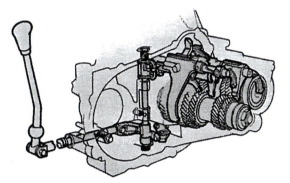

Figure 6.11 FWD gearbox selector mechanism

Final drive and differential

The final drive and differential systems were covered in the level 2 book but a bit of revision is always useful! On cars with rear wheel drive, the final drive transmits power from the propshaft to axle shafts connected to the road wheels. The propeller shaft and the axle are at right angles to each other. The final drive must therefore carry power through this angle to drive the wheels. In addition to carrying the power to the drive wheels, the differential divides the torque evenly between those two wheels even though the wheels may be turning at different speeds when cornering. The final drive gears provide a fixed gear reduction from the gearbox to the road wheels.

The crown wheel and pinion are types of bevel gears because they mesh at right angles to each other. They carry power through a right angle to the drive wheels. The crown wheel is driven by the pinion, which receives power from the propeller shaft. In most cars, the ring gear is two to four times as large as the pinion. This means that these gears reduce the speed from the propeller shaft and increase the torque. The reduction in the final drive multiplies the reduction that has already taken place in the

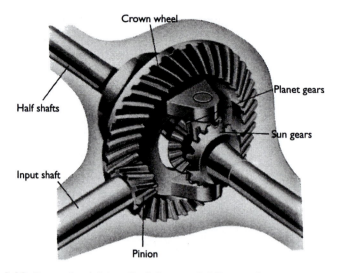

Figure 6.12 Rear wheel drive, final drive and differential

transmission. This system is shown as Figure 6.12. The crown wheel gear is known as a 'hypoid-type' named after the way the teeth are cut. As well as quiet operation, this allows the pinion to be set lower than the crown wheel centre. This saves space in the vehicle as a smaller transmission tunnel can be used.

Most cars now have a transverse engine, which drives the front wheels. The power of the engine, therefore, does not have to be carried through a right angle to the drive wheels. The final drive contains ordinary reducing gears rather than bevel gears (Figure 6.13).

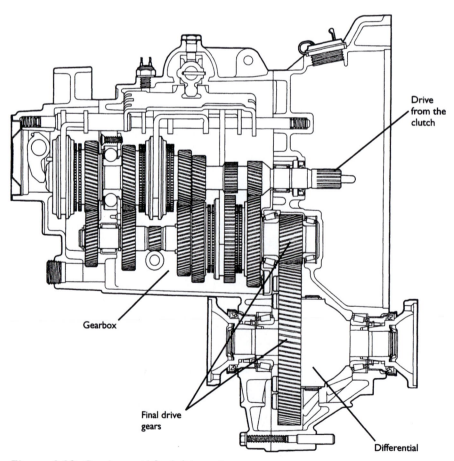

Figure 6.13 Gearbox and final drive unit

KEY WORDS

- Synchromesh
- Detent
- Interlock
- Ratio
- Baulk ring
- Multiplate
- Overdrive
- Centre differential

The differential is a set of gears that divide the torque evenly between the two drive wheels. The differential also allows one wheel to rotate faster than the other when necessary. When a car goes around a corner, the outside drive wheel travels further than the inside one. The outside wheel must therefore rotate faster than the inside one to cover the greater distance in the same time.

Some higher performance vehicles use limited slip differentials. An electronic-hydraulic controlled unit is shown in Figure 6.14. The clutch plates are connected to the two output shafts and, if controlled, will in turn control the amount of slip. This can be used to counteract the effect of one wheel losing traction when high power is applied.

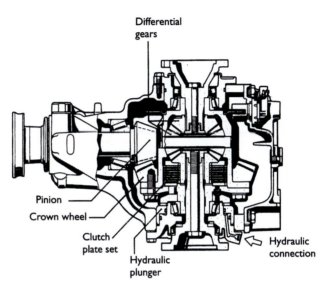

Figure 6.14 Final drive with electronic-hydraulic limited-slip differential

Differential locks are used on many off-road type vehicles. A simple dog clutch or similar device prevents the differential action. This allows far better traction on slippery surfaces.

Overdrive

On earlier vehicles, a four-speed gearbox was the norm. Further improvements in operation could be gained by fitting an overdrive. This was mounted on the output of the gearbox (RWD). In fourth gear, the drive ratio is usually 1:1. Overdrive would allow the output to rotate faster than the input, hence the name. Most gearboxes now incorporate a fifth gear (refer back to Figures 6.7 and 6.8) which is effectively an overdrive but does not form a separate unit.

Four-wheel drive systems

Four-wheel drive provides good traction on rough or slippery surfaces. Many cars are now available with four-wheel drive. In some vehicles, the driver can switch between four-wheel drive and two-wheel drive. A vehicle with four-wheel drive delivers power to all four wheels. A transfer box is used to distribute the power between the front and rear wheels. Figure 6.15 shows the layout of the system used on the Ford Maverick and Figure 6.16 shows a similar system in more detail. Note the extra components compared to a normal two-wheel drive system:

- transfer gearbox – to provide an extra drive output
- differential on each axle – to allow cornering speed variations
- centre differential – to prevent wind-up between the front and rear axles
- extra drive shafts – to supply drive to the extra axle.

Figure 6.15 Ford Maverick 4WD transmission

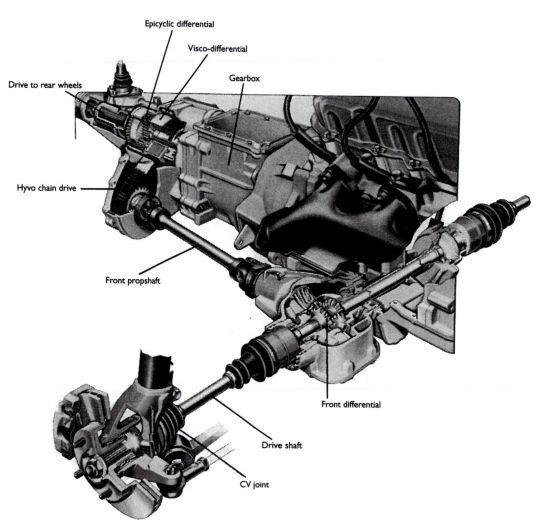

Figure 6.16 Details of 4WD transmission

The transfer gearbox on some vehicles may also contain extra reduction gears for low ratio drive.

One problem to overcome however with 4WD is that, if three differentials are used then the chance of one wheel slipping actually increases. This is because the drive will always be transferred to the wheel with least traction – like running a 2WD car with one driving wheel jacked up. To overcome this problem, and take advantage of the extra traction available, a viscous coupling is combined with an epicyclic gear train to form the centre differential.

The drive can now be distributed proportionally. A typical value is about 35% to the front and 65% to the rear wheels. However, the viscous clutch coupling shown in Figure 6.16 acts so that if, for example, the rear wheels start to spin, the greater difference in speed across the coupling will cause more friction and hence more to occur through the coupling. This tends to act so that the drive is automatically distributed to the most effective driving axle. A 'Hyvo' or silent chain drive is often used to drive from the transfer box.

LEARNING TASKS

→ Look back at the key words. Explain each one to a friend, and/or write out a short description to keep as evidence.

→ Make a simple sketch to show a gearbox detent mechanism.

→ Examine a real manual gearbox and note the components in detail.

→ Write a short explanation about how a differential operates.

---

3 Automatic transmission

Introduction

An automatic gearbox contains special devices that automatically provide various gear ratios as they are needed. Most automatic gearboxes have three or four forward gears and reverse. Instead of a gearstick, the driver moves a lever called a selector. Most automatic gearboxes now have selector positions for park, neutral, reverse, drive, 2 and 1 (or 3, 2 and 1 in some cases). The engine will only start if the selector is in either the park or neutral position. In park, the drive shaft is locked so that the drive wheels cannot move. It is now quite common, when the engine is running, to only be able to move the selector out of park if you are pressing the brake pedal. This is a very good safety feature as it prevents sudden movement of the vehicle.

For ordinary driving, the driver moves the selector to the drive position. The transmission starts out in the lowest gear and automatically shifts into higher gears as the car picks up speed. The driver can use the lower positions of the gearbox for going up or down steep hills or driving through mud or snow. When in position 3, 2, or 1, the gearbox will not change above the lowest gear specified. Figure 6.17 shows a very common type of automatic gearbox used on rear wheel drive vehicles. This box contains three forward gears and a reverse.

Fluid flywheel and torque converter operation

A fluid flywheel consists of an impeller and a turbine immersed in oil to transmit drive from the engine to the gearbox. The engine driven impeller faces the turbine, which is connected to the gearbox. Each part is bowl shaped and contains a number of vanes. They are both a little like half of a

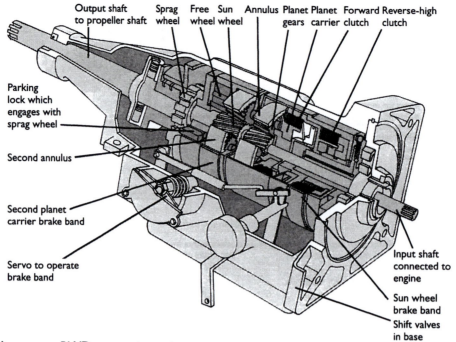

Figure 6.17 A common RWD automatic gearbox

hollowed out orange facing each other. When the engine is running at idle speed, oil is flung from the impeller into the turbine but not with enough force to turn the turbine. As engine speed increases, so does the energy of the oil. This increasing force begins to move the turbine and, hence, the vehicle. The oil gives up its energy to the turbine and then recirculates into the impeller at the centre, starting the cycle over again. As the vehicle accelerates, the difference in speed between the impeller and turbine reduces until the slip is about 2%. A problem, however, with a basic fluid flywheel is that it is slow to react when the vehicle is moving off from rest. This can be improved by fitting a reactor or stator between the impeller and turbine. It is this device that we now know as a torque converter.

The torque converter is a device that almost all automatic transmissions now use. It delivers power from the engine to the gearbox like a basic fluid flywheel, but also increases the torque when the car begins to move. Similar to a fluid flywheel, the torque converter resembles a large doughnut sliced in half. One half, called the pump impeller, is bolted to the drive plate or flywheel. The other half, called the turbine, is connected to the gearbox-input shaft. Each half is lined with vanes or blades. The pump and the turbine face each other in a case filled with oil. A bladed wheel called a stator is fitted between them.

The engine causes the pump to rotate and throw oil against the vanes of the turbine. The force of the oil makes the turbine rotate and send power to the transmission. After striking the turbine vanes, the oil passes through the stator and returns to the pump. When the pump reaches a specific rate of rotation, a reaction between the oil and the stator increases the torque. In a fluid flywheel, oil returning to the impeller tends to slow it down. In a torque converter, the stator or reactor diverts the oil towards the centre of the impeller for extra thrust. Figure 6.18 shows the operation of a fluid flywheel and a torque converter.

When the engine is running slowly, the oil may not have enough force to rotate the turbine. But when the driver presses the accelerator pedal, the engine runs faster and so does the impeller. The action of the impeller increases the force of the oil. This force gradually becomes strong enough to

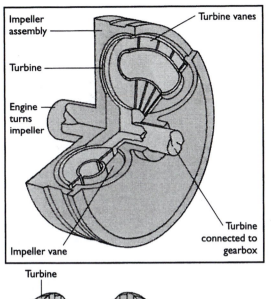

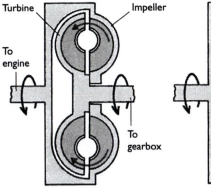

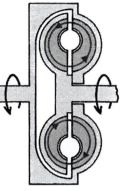

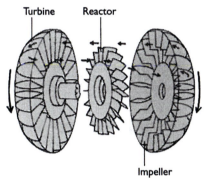

At low engine speeds, centrifugal force throws oil from the impeller to the turbine and rotates the turbine slowly

At high engine speeds, the oil moves more quickly, forming a fluid bond between the two halves with very little slip

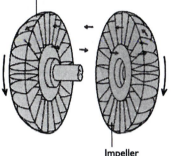

The fluid flywheel is slow to react when moving from a standstill. The torque converter has a centre reactor which re-directs oil flow at low speeds, giving increased 'bite' when moving off

Figure 6.18 Fluid flywheel and torque converter operation

rotate the turbine and moves the vehicle. A torque converter can double the applied torque when moving off from rest. As engine speed increases, the torque multiplication tapers off until at cruising speed there is no increase in torque. The reactor or stator then freewheels on its one-way clutch at the same speed as the turbine.

The fluid flywheel action reduces efficiency because the pump tends to rotate faster than the turbine. In other words, some slip will occur (about 2% as suggested previously). To improve efficiency, many transmissions now include a lock-up clutch. When the pump reaches a specific rate of rotation, this clutch locks the pump and turbine together, allowing them to rotate as one.

Epicyclic gearbox operation

Epicyclic gears are a special set of gears that are part of most automatic gearboxes. They consist of three elements:

- a sun gear, located in the centre
- the carrier that holds two, three, or four planet gears, which mesh with the sun gear and revolve around it
- an internal gear or annulus – a ring with internal teeth, it surrounds the planet gears and meshes with them.

Any part of a set of planetary gears can be held stationary or locked to one of the others. This will produce different gear ratios (Figure 6.19). Most of the automatic gearboxes of the type shown in Figure 6.17 have two sets of planetary gears that are arranged in line. This provides the necessary number of gear ratios. The appropriate elements in the gear train are held stationary by a system of hydraulically operated brake bands and clutches.

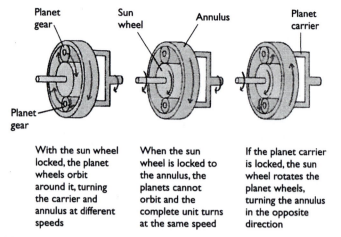

| With the sun wheel locked, the planet wheels orbit around it, turning the carrier and annulus at different speeds | When the sun wheel is locked to the annulus, the planets cannot orbit and the complete unit turns at the same speed | If the planet carrier is locked, the sun wheel rotates the planet wheels, turning the annulus in the opposite direction |

Figure 6.19 Epicyclic gear set

These are worked by a series of hydraulically operated valves in the lower part of the gearbox. Figure 6.17 is a three speed box; an extra set of gears is required for four speeds. Oil pressure to operate the clutches and brake bands is supplied by a pump. The supply for this is the oil in the sump of the gearbox. Figure 6.20 shows how the three forward gears and one reverse gear are achieved from two sets of epicyclic gears.

Unless the driver moves the gear selector to operate the valves, automatic gear changes are made depending on just two factors:

- throttle opening – a cable is connected from the throttle to the gearbox
- road speed – when the vehicle reaches a set speed, a governor allows pump pressure to take over from the throttle.

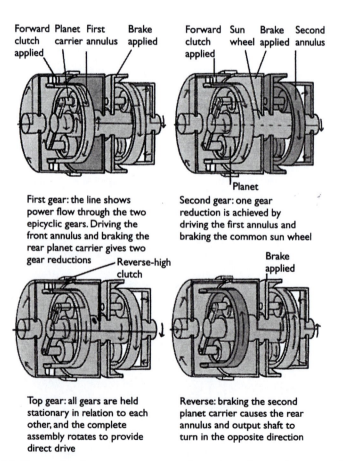

First gear: the line shows power flow through the two epicyclic gears. Driving the front annulus and braking the rear planet carrier gives two gear reductions

Second gear: one gear reduction is achieved by driving the first annulus and braking the common sun wheel

Top gear: all gears are held stationary in relation to each other, and the complete assembly rotates to provide direct drive

Reverse: braking the second planet carrier causes the rear annulus and output shaft to turn in the opposite direction

Figure 6.20 Two epicyclic gear sets can produce three forward gears and reverse

KEY WORDS

■ CVT / CTX
■ ECAT
■ Epicyclic
■ Fluid flywheel
■ Torque converter
■ Kick down
■ Slip
■ Electronic clutch

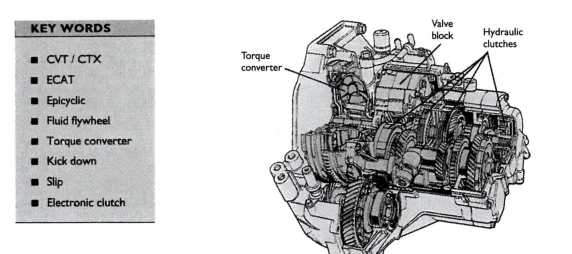

Figure 6.21 Automatic gearbox and torque converter – using conventional gears and hydraulic clutches

The cable from the throttle also allows a facility known as 'kick down'. This allows the driver to change down a gear, such as for overtaking, by pressing the throttle all the way down.

Many modern automatic gearboxes now use gears the same as in manual boxes. The changing of ratios is similar to the manual operation except that hydraulic clutches and valves are used. This is covered further in the case study on page 137. Figure 6.21 shows an example of this system.

Constantly variable transmission

Figure 6.22 shows the Ford CTX (Constantly variable TransaXle) transmission. This kind of automatic transmission called a continuously variable transmission uses two pairs of cone-shaped pulleys connected by a

Figure 6.22 Ford CTX transmission

metal belt. The key to this system is the high friction drive belt. The belt is made from high performance steel and transmits drive by thrust rather than tension. The ratio of the rotations, often called the gear ratio, is determined by how far the belt rides from the centres of the pulleys. The transmission can produce an unlimited number of ratios. As the car changes speed, the ratio is continuously adjusted. Cars with this system are said to use fuel more efficiently than cars with set gear ratios. Within the gearbox, hydraulic control is used to move the pulleys and hence change the drive ratio. An epicyclic gear set is used to provide a reverse gear as well as a fixed ratio.

Electronic control of transmission

The main aim of electronically controlled automatic transmission (ECAT) is to improve on conventional automatic transmission in the following ways:

- gear changes should be smoother and quieter
- improved performance
- reduced fuel consumption
- reduction of characteristic changes over system life
- increased reliability.

The important points to remember are that gear changes and lock-up of the torque converter are controlled by hydraulic pressure. In an ECAT system, electrically controlled solenoid valves can influence this hydraulic pressure. Most ECAT systems now have a transmission ECU that is in communication with the engine control ECU. Control of gear shift and torque converter lockup.

With an ECAT system, the actual point of gearshift is determined from pre-programmed memory within the ECU. Data from other sensors is also taken into consideration. Actual gearshifts are initiated by changes in hydraulic pressure, which is controlled by solenoid valves.

The two main control functions of this system are hydraulic pressure and engine torque. A temporary reduction in engine torque during gear shifting allows smooth operation. This is because the peaks of gearbox output torque, which cause the characteristic surge during gear changes on conventional automatics, is suppressed. Because of these control functions, smooth gearshifts are possible and – due to the learning ability of some ECU's – the characteristics remain throughout the life of the system.

The ability to lock-up the torque converter has been used for some time, even on vehicles with more conventional automatic transmission. This gives better fuel economy, quietness and improved driveability. Lock-up is carried out using a hydraulic valve, which can be operated gradually to produce a smooth transition. The timing of lock-up is determined from ECU memory in terms of the vehicle's speed and acceleration.

Semi-automatic transmission

A number of different types of semi-automatic transmission are either in use or under development. An interesting system is the electronically controlled clutch. This technology is a spin-off from Formula 1 and top rally cars. The electronic clutch was developed for these racing vehicles to improve the get-away performance and speed of gear changes.

For production vehicles, a system has been developed which can interpret the drivers intention. With greater throttle openings, the clutch operation changes to prevent abuse and drive line damage. Electrical control of the clutch release bearing position is by a solenoid actuator or electric motor with worm gear reduction. This allows the time to reach the ideal take off position to be reduced, and the ability of the clutch to transmit torque to be improved.

This technique has now been developed to the stage where it will operate the clutch automatically as the stick is moved to change gear. Sensors are fitted to monitor engine and road speed as well as gear position. The Ferrari Mondiale uses a system similar to the one described here. The system used on a variant of the 97/8 Renault Clio is also similar in operation in that it has a normal gearbox but no clutch pedal. Saab is also well known for its use of semi-automatic transmission.

LEARNING TASKS

➡ Look back at the key words. Explain each one to a friend, and/or write out a short description to keep as evidence.

➡ Make a simple sketch to show an epicyclic gear train.

➡ Examine a real automatic box and note the components in detail.

➡ Write a short explanation about why different drive positions are available from automatic gearboxes.

4 Rover 825 automatic transmission – case study

Introduction   The automatic transmission is mounted in-line with the engine and comprises: a three-element torque converter, four forward speeds and reverse. Figure 6.23 shows the main components of this gearbox. The

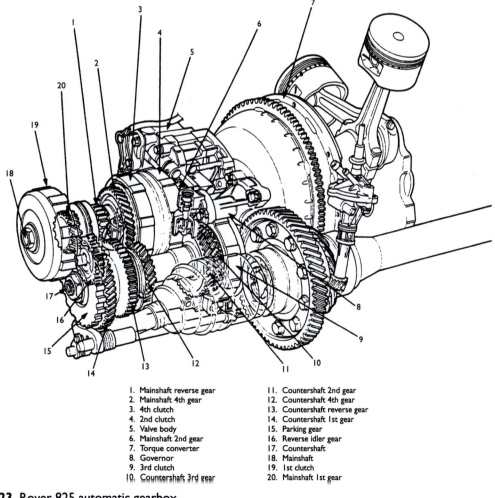

1. Mainshaft reverse gear	11. Countershaft 2nd gear
2. Mainshaft 4th gear	12. Countershaft 4th gear
3. 4th clutch	13. Countershaft reverse gear
4. 2nd clutch	14. Countershaft 1st gear
5. Valve body	15. Parking gear
6. Mainshaft 2nd gear	16. Reverse idler gear
7. Torque converter	17. Countershaft
8. Governor	18. Mainshaft
9. 3rd clutch	19. 1st clutch
10. Countershaft 3rd gear	20. Mainshaft 1st gear

Figure 6.23 Rover 825 automatic gearbox

gearbox contains two shafts, the mainshaft and the countershaft. This is an interesting system because the gearbox is similar to a manual system but it is operated automatically. The shafts run parallel to each other and carry the following components:

Mainshaft	1st gear and 1st gear clutch
	2nd gear, 4th gear and 2nd/4th clutch
	3rd gear (integral with the mainshaft)
Countershaft	Parking gear, reverse gear
	1st gear with internally mounted one-way clutch
	2nd gear and 4th gear
	3rd gear and 3rd gear clutch

All gears run on needle roller bearings except 2nd gear on the countershaft, which is splined to the shaft. All gears (except reverse) on the mainshaft are in constant mesh with their corresponding gear on the countershaft. Drive is transmitted from the mainshaft to the countershaft by the engagement of clutches or reverse gear servo valve.

Torque converter   The torque converter is a three-element type incorporating an electronic/hydraulic controlled lock-up clutch. In 'D' (Drive) range, the clutch will, under certain conditions, lock the turbine to the impeller. The stator is splined to a stator shaft; this applies an additional force on the regulator valve during acceleration, thus increasing line pressure in the hydraulic system and providing additional clamping pressure on the gear clutch plates.

The three elements of the torque converter are:

■   impeller – driven by the engine
■   stator – mounted on a one-way spring clutch between the impeller and the turbine
■   turbine – drives the gearbox via the input shaft.

In a conventional type of torque converter, there is always an inherent amount of 'slip' between the impeller and turbine. This 'slip' causes an increase in fuel consumption, especially during high-speed cruising. This is eliminated in this gearbox by means of a lock-up mechanism in the torque converter.

The mechanism is made up of a friction lining and a piston. With the transmission operating in the 'D' (Drive) range, 2nd speed and above, the piston, operating under hydraulic pressure, applies the friction lining to the impeller, thereby eliminating any slip between the impeller and the turbine. The system is controlled by the transmission governor and throttle pressure, and by a lock-up control solenoid valve operated by the engine management ECU. Operation of the lock-up system is also dependent on coolant temperature, vehicle speed and throttle opening.

When throttle opening, vehicle speed, coolant temperature and governor pressure exceed certain limits, oil between the converter cover and lock-up piston is discharged, converter pressure enters through the inlet and exerts pressure on the lock-up piston, engaging the lining with the impeller. The effect is to by-pass the converter, thereby placing the torque converter in direct drive.

When throttle opening, vehicle speed, coolant temperature and governor pressure fall below certain limits, converter pressure enters via the inlet and moves the lock-up piston away from the impeller, disengaging the lining. The lock-up mechanism is now released and the impeller and turbine can rotate at different speeds.

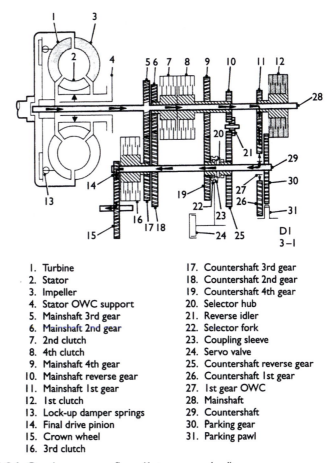

1. Turbine
2. Stator
3. Impeller
4. Stator OWC support
5. Mainshaft 3rd gear
6. Mainshaft 2nd gear
7. 2nd clutch
8. 4th clutch
9. Mainshaft 4th gear
10. Mainshaft reverse gear
11. Mainshaft 1st gear
12. 1st clutch
13. Lock-up damper springs
14. Final drive pinion
15. Crown wheel
16. 3rd clutch

17. Countershaft 3rd gear
18. Countershaft 2nd gear
19. Countershaft 4th gear
20. Selector hub
21. Reverse idler
22. Selector fork
23. Coupling sleeve
24. Servo valve
25. Countershaft reverse gear
26. Countershaft 1st gear
27. 1st gear OWC
28. Mainshaft
29. Countershaft
30. Parking gear
31. Parking pawl

Figure 6.24 Gearbox power flow (1st gear marked)

Mechanical operation and power flows	The following table describes the power flows through the gearbox; refer to Figure 6.24 to follow them through. In each case, power enters the transmission from the torque converter via the mainshaft.

KEY WORDS

■ Power flow
■ Lock up

Position	Power flow	Gear ratio
'D1' and '3–1'	The 1st clutch is applied by hydraulic pressure and locks the mainshaft 1st gear to the mainshaft. The mainshaft 1st gear drives the countershaft 1st gear and power is transmitted via the one-way clutch onto the countershaft and final drive gear	2.647:1
'D2' and '3–2'	The 2nd clutch is applied by hydraulic pressure and locks the mainshaft 2nd gear to the mainshaft. The mainshaft 2nd gear rotates, causing the countershaft 2nd gear to transmit the power to the countershaft and thereby to the final drive gear. The 1st clutch remains on in 2nd gear and, to prevent the transmission locking up, the one-way clutch (OWC) free wheels	1.555:1
'D3' and '3–3'	The 3rd clutch is applied by hydraulic pressure and locks the countershaft 3rd gear to the countershaft. The mainshaft 3rd gear rotates causing the countershaft 3rd gear to transmit the power to the countershaft and, thereby, to the final drive gear. The 1st clutch remains on in 3rd gear and, to prevent the transmission locking-up, the one-way clutch free wheels	0.971:1

Position	Power flow	Gear ratio
'D4'	The 4th clutch is applied by hydraulic pressure and locks the mainshaft 4th gear to the mainshaft. The mainshaft 4th gear rotates, causing the countershaft 4th gear to transmit the power to the coupling sleeve and selector hub. The selector hub being splined to the countershaft transmits the power to the final drive gear. The 1st clutch remains on in 4th gear and, to prevent the transmission locking-up, the one-way clutch free wheels	0.682:1
'2'	The 2nd clutch is applied by hydraulic pressure and locks the mainshaft 2nd gear to the mainshaft. The mainshaft 2nd gear rotates, causing the countershaft 2nd gear to transmit the power to the countershaft and thereby to the final drive gear	1.555:1
'R'	The 4th clutch is applied by hydraulic pressure and locks the mainshaft 4th gear to the mainshaft. Hydraulic pressure acting upon the servo valve moves the coupling sleeve into engagement with the countershaft reverse gear, thereby locking the gear to the countershaft. The mainshaft reverse gear is part of the mainshaft, therefore 4th gear rotates. The reverse gear and countershaft reverse gear now transmits the power to the final drive gear via the countershaft. Because the power is transmitted by the reverse idler gear, the countershaft rotates in the opposite direction	1.904:1

Hydraulic control system   The hydraulic control system consists of a main valve body onto which is mounted the regulator valve body and the secondary valve body (Figure 6.25). The detailed operation of the hydraulic system is beyond the scope of this book. However Figure 6.26 should serve to give you a basic idea of how

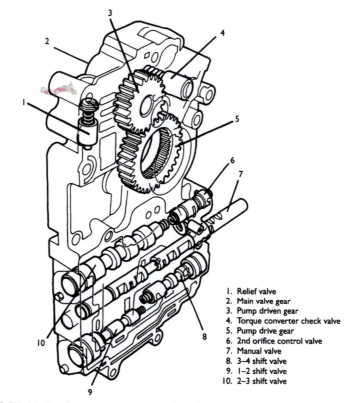

1. Relief valve
2. Main valve gear
3. Pump driven gear
4. Torque converter check valve
5. Pump drive gear
6. 2nd orifice control valve
7. Manual valve
8. 3–4 shift valve
9. 1–2 shift valve
10. 2–3 shift valve

Figure 6.25 Hydraulic control valves and body

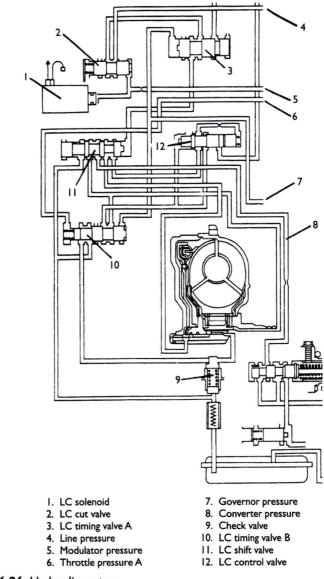

1. LC solenoid
2. LC cut valve
3. LC timing valve A
4. Line pressure
5. Modulator pressure
6. Throttle pressure A
7. Governor pressure
8. Converter pressure
9. Check valve
10. LC timing valve B
11. LC shift valve
12. LC control valve

Figure 6.26 Hydraulic system

the system operates. The different valves are controlled either by fluid
pressure, electrical signals or manual operation of the selector. The valves in
turn are able to control the clutches to lock appropriate gears onto their
shafts, so transmitting drive.

LEARNING TASKS

➡ Look back at the key words. Explain each one to a friend, and/or write out a
short description to keep as evidence.

➡ Make a simple sketch to show power flow in 2nd, 3rd and 4th gear.

7 Suspension

1 Introduction

The basic functions of automobile suspension are as follows:

■ allowing the wheels to follow ground relief without transmitting excessive strain to the body

■ ensuring the wheels remain in contact with the ground

■ minimising movements of the suspended assembly, and particularly the passenger compartment.

Suspension involves the combined action of springs and dampers. In order to absorb or reduce shocks and vibration caused by surface irregularities, a spring or 'elastic' assembly must be fitted between the wheel and the body. Figure 7.1 is a vehicle cutaway to show suspension components.

Figure 7.1 A Fiesta Si showing suspension and other components

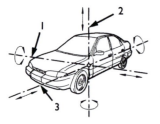

1. Transverse axis
2. Vertical axis
3. Longitudinal axis

Figure 7.2 Vehicle oscillations

To ensure that these shocks and vibrations are absorbed effectively, the suspension must be highly flexible. However, bumps tend to make the wheels jump off the road surface and then to rebound. This bump and rebound has a negative effect on road holding. The softer the springs are, the more pronounced this movement will be. It is the damper's role to control the rocking or bouncing movement. The damper has two tasks:

■ overcome the rocking motion of the body

■ keep the wheels in contact with the ground.

The spring-damper assembly forms the basis of all suspension systems, and the characteristics of this assembly determine the quality of the suspension system. Figure 7.2 shows the effect of vehicle oscillations.

Terminology

Spring	The device that takes the road shocks. It must have elastic properties. The most common spring is the coil spring but many others are used, including torsion bar, rubber, gas and leaf springs
Damper	Reduces oscillation of the suspension. The damper converts the mechanical energy in the spring into heat. This is most often done by causing fluid to pass through small holes very quickly
Longitudinal axis	The long axis of the vehicle in line with the direction of travel
Lateral axis	The sideways or transverse axis of the vehicle at ninety degrees to the longitudinal axis, or parallel with the wheel axles
Progressive	In relation to springing, this means that the more a spring is compressed then the greater the force needed to compress it further (Figure 7.3)

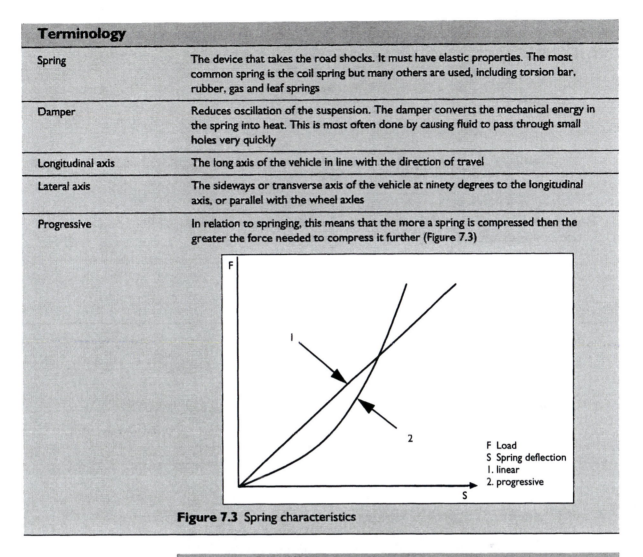

F Load
S Spring deflection
1. linear
2. progressive

Figure 7.3 Spring characteristics

LEARNING TASKS

➡ Look back at the key words. Explain each one to a friend, and/or write out a short description to keep as evidence.

➡ Make a simple sketch to show the forces acting on a vehicle as it is being driven over rough ground.

---

2 Suspension systems

Introduction

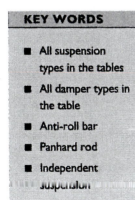

KEY WORDS

- All suspension types in the tables
- All damper types in the table
- Anti-roll bar
- Panhard rod
- Independent suspension

Traditionally beam axles were used to support the wheels. This is still the case on most commercial vehicles because the design is very strong. Having the left and right side wheels connected together, however, means that a bump on one side of the vehicle will affect the opposite wheel as well.

Independent suspension is achieved by using systems of arms and links. Pressed steel or tube sections are used, often shaped in the form of a triangle for strength. One corner of the triangle holds the wheel and the other two corners pivot on the body. If this pivot axis is parallel to the longitudinal axis of the vehicle, they are called transverse links, transverse arms or wishbones. If the axis runs at right angles to the direction of travel, they are called leading or trailing arms. If the axis is at an angle, they are called semi-trailing arms. As the wheel moves, these arms should be arranged so that its angle should remain within tight limits.

Road holding is an important aspect of suspension. It is influenced by the following factors:

- suspension system layout
- centre of gravity position
- weight distribution between axles
- type of suspension
- tyres
- body design.

Requirements of the suspension system

The suspension system is the link between the vehicle body and the wheels. Its purpose can be summarised as follows:

- locate the wheels
- allow wheels to move up and down
- allow wheels to steer
- absorb driving forces
- maintain the wheels in contact with the road
- distribute the weight of the vehicle to the wheels
- reduce road noise from the tyres
- produce a compromise between comfort and safety
- keep the vehicle weight as low as possible – in particular the unsprung mass.

Sprung and unsprung mass

Unsprung mass is defined as the mass of the suspension component, the wheels and the springs. However, only 50% of the spring mass and the moving suspension arms are included because they form part of the link between the sprung and unsprung masses.

It is beneficial to have the unsprung mass as small as possible in comparison with the sprung mass (main vehicle mass). This is so that when the vehicle hits a bump, the movement of the suspension will have only a small effect on the main part of the vehicle. Figure 7.4 shows a representation of this effect. The overall result is therefore improved ride comfort.

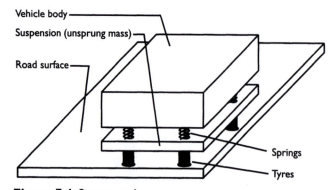

Figure 7.4 Sprung and unsprung mass

Front axle suspension systems

With front axle suspension systems, as with most design aspects of the vehicle, compromise often has to be reached between performance, body styling and cost. The following table compares most of the main front axle suspension systems:

Name	Description	Advantages	Disadvantages
Double transverse arms (Figure 7.5)	Independently suspended wheels located by two arms perpendicular to direction of travel. The arms support axles	Low bonnet line. Only slight changes of track and camber with suspension movements	A large number of pivot points is required. High production costs

1. Upper transverse arm
2. Lower transverse arm

Figure 7.5 Double transverse arms

Name	Description	Advantages	Disadvantages
Transverse arms with leaf spring (Figure 7.6)	A traverse arm and a leaf spring locate the wheel	The spring can act as an anti roll bar, hence low cost	Harsh response when lightly loaded. Major changes of camber as vehicle is loaded

Transverse leaf spring mounted above

Transverse leaf spring mounted below and supported at two points (in this way it can act like an anti-roll bar)

Figure 7.6 Transverse arm with leaf spring

Name	Description	Advantages	Disadvantages
Transverse arm with McPherson strut (Figures 7.7 and 7.8)	A combination of the spring, damper, wheel hub, steering arm and axle joints in one unit	Only slight changes in track and camber with suspension movement. Forces on the joints are reduced because of the long strut	The body must be strengthened around the upper mounting. A low bonnet line is difficult

1. Suspension strut
2. Transverse arm

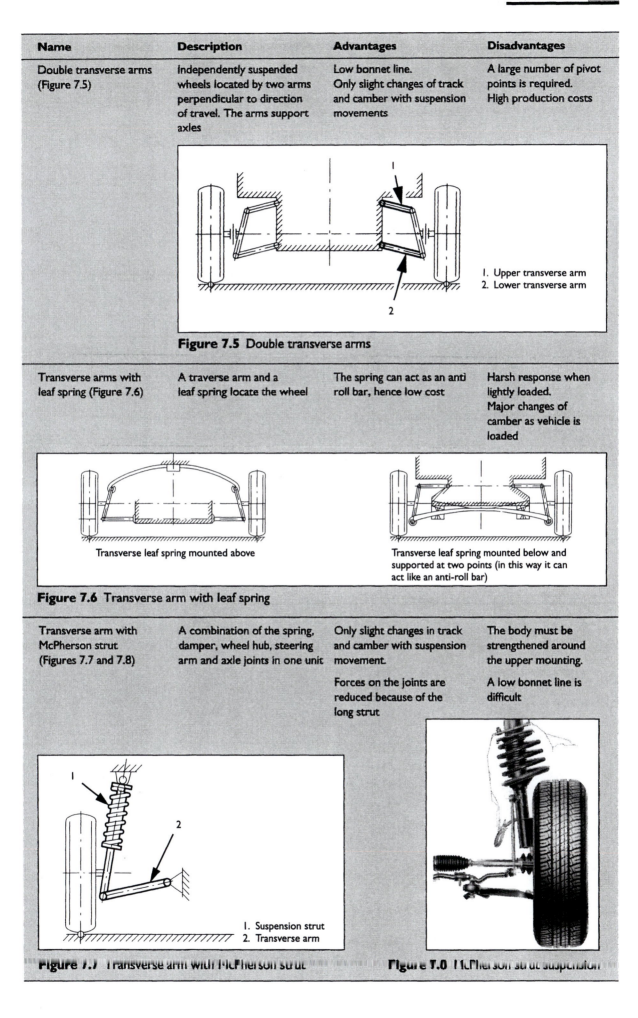

Figure 7.7 Transverse arm with McPherson strut

Figure 7.8 McPherson strut suspension

Name	Description	Advantages	Disadvantages
Double trailing arms (Figure 7.9)	Two trailing arms support the stub axle. These can act on torsion bars often formed a a single assembly	No change in castor, camber or track with suspension movement Can be assembled and adjusted off the vehicle	Lots of space is required at the front of the vehicle Expensive to produce. Acceleration and braking cause pitching movements which, in turn, changes the wheel base

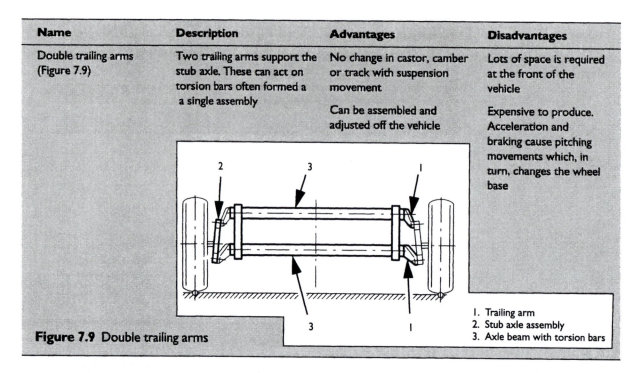

1. Trailing arm
2. Stub axle assembly
3. Axle beam with torsion bars

Figure 7.9 Double trailing arms

Rear axle suspension systems The following table compares most of the main rear axle suspension systems:

Name	Description	Advantages	Disadvantages
Rigid axle with leaf springs (Figure 7.10)	The final drive, differential and axle shafts are all one unit	Rear track remains constant reducing tyre wear. Good directional stability because no camber change causes body roll on corners. Low cost. Strong design for load carrying	High unsprung mass The interaction of the wheels causes lateral movement, reducing tyre adhesion when the suspension is compressed on one side

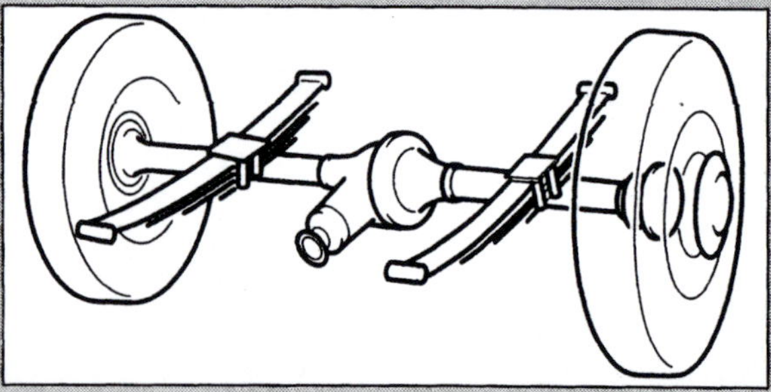

Figure 7.10 Rigid axle with leaf springs

Name	Description	Advantages	Disadvantages
Rigid axle with A-bracket (Figure 7.11)	Solid axle with coil springs and a central joint supports the axle on the body	Rear of the vehicle pulls down on braking, which stabilises the vehicle	High cost. Large unsprung mass

A-bracket

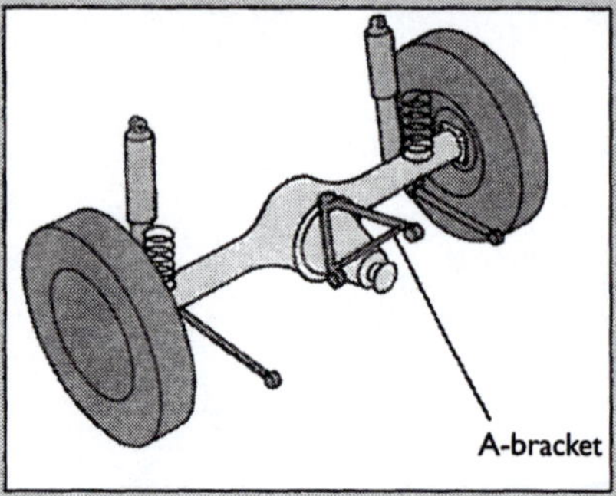

Figure 7.11 Rear axle with A-bracket

Name	Description	Advantages	Disadvantages
Rigid axle with compression/tension struts (Figure 7.12)	Coil springs provide the springing and the axle is located by struts	Suspension extension is reduced when braking or accelerating. The springs are isolated from these forces	High loads on the welded joints. High weight overall. Large unsprung mass

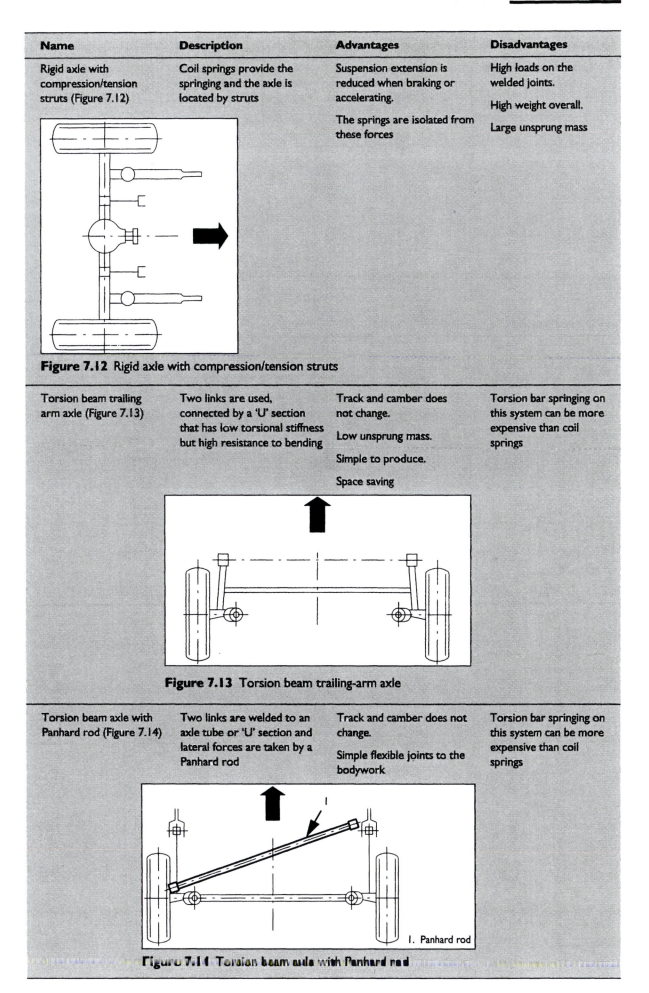

Figure 7.12 Rigid axle with compression/tension struts

Name	Description	Advantages	Disadvantages
Torsion beam trailing arm axle (Figure 7.13)	Two links are used, connected by a 'U' section that has low torsional stiffness but high resistance to bending	Track and camber does not change. Low unsprung mass. Simple to produce. Space saving	Torsion bar springing on this system can be more expensive than coil springs

Figure 7.13 Torsion beam trailing-arm axle

Name	Description	Advantages	Disadvantages
Torsion beam axle with Panhard rod (Figure 7.14)	Two links are welded to an axle tube or 'U' section and lateral forces are taken by a Panhard rod	Track and camber does not change. Simple flexible joints to the bodywork	Torsion bar springing on this system can be more expensive than coil springs

1. Panhard rod

Figure 7.14 Torsion beam axle with Panhard rod

Name	Description	Advantages	Disadvantages
Trailing arms (Figure 7.15)	The pivot axis of the trailing arms is at ninety degrees to the direction of vehicle travel	When braking, the rear of the vehicle pulls down, giving stable handling. Track and camber does not change. Space saving	Slight change of wheel base when the suspension is compressed

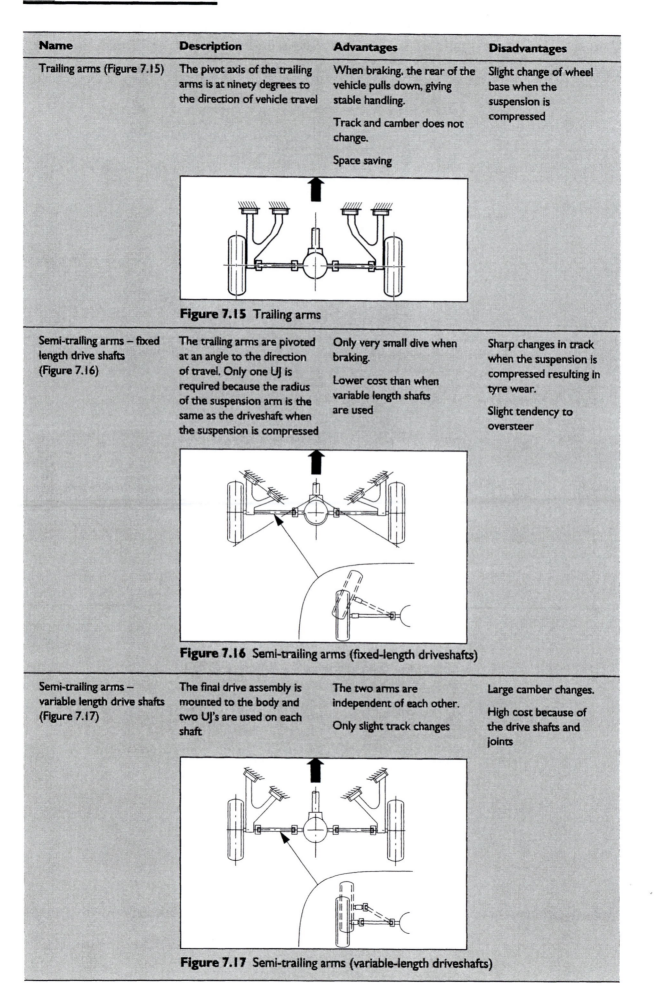

Figure 7.15 Trailing arms

Name	Description	Advantages	Disadvantages
Semi-trailing arms – fixed length drive shafts (Figure 7.16)	The trailing arms are pivoted at an angle to the direction of travel. Only one UJ is required because the radius of the suspension arm is the same as the driveshaft when the suspension is compressed	Only very small dive when braking. Lower cost than when variable length shafts are used	Sharp changes in track when the suspension is compressed resulting in tyre wear. Slight tendency to oversteer

Figure 7.16 Semi-trailing arms (fixed-length driveshafts)

Name	Description	Advantages	Disadvantages
Semi-trailing arms – variable length drive shafts (Figure 7.17)	The final drive assembly is mounted to the body and two UJ's are used on each shaft	The two arms are independent of each other. Only slight track changes	Large camber changes. High cost because of the drive shafts and joints

Figure 7.17 Semi-trailing arms (variable-length driveshafts)

Anti-roll bar The main purpose of an anti-roll bar (Figure 7.18) is to reduce body roll on corners. The anti-roll bar can be thought of as a torsion bar. The centre is pivoted on the body and each end bends to make connection with the suspension/wheel assembly. When the suspension is compressed on both sides, the anti-roll bar has no effect because it pivots on its mountings.

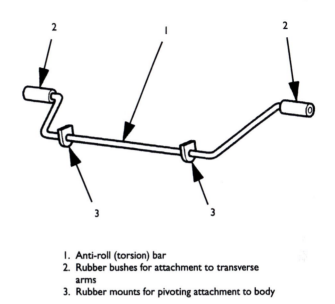

1. Anti-roll (torsion) bar
2. Rubber bushes for attachment to transverse arms
3. Rubber mounts for pivoting attachment to body

Figure 7.18 Anti-roll bar and mountings

As the suspension is compressed on just one side, a twisting force is exerted on the anti-roll bar. The anti-roll bar is now under torsional load. Part of this load is transmitted to the opposite wheel, pulling it upwards. This reduces the amount of body roll on corners. The disadvantages are that some of the 'independence' is lost and the overall ride is harsher. Anti-roll bars can be fitted to both front and rear axles.

Panhard rod Named after a French engineer, the Panhard rod links the rear axle to the body. It is pivoted at each end to allow movement. The rod (Figure 7.14) takes up lateral forces between the axle and body, thus removing load from the radius arms. The radius arms, therefore, have only to transmit longitudinal forces.

Springs The requirements of the springs can be summarised as follows:

- absorb road shocks from uneven surfaces
- control ground clearance and ride height
- ensure good tyre adhesion
- support the weight of the vehicle
- transmit gravity forces to the wheels.

There are a number of different types of spring in use on modern light vehicles. The following table lists these together with their main features:

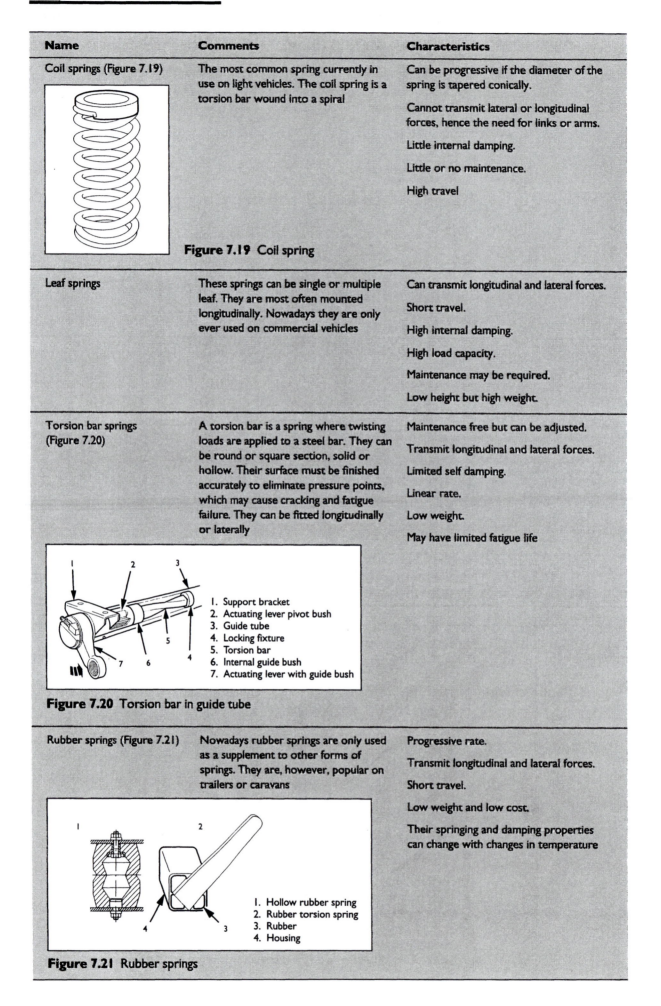

Name	Comments	Characteristics
Coil springs (Figure 7.19)	The most common spring currently in use on light vehicles. The coil spring is a torsion bar wound into a spiral	Can be progressive if the diameter of the spring is tapered conically. Cannot transmit lateral or longitudinal forces, hence the need for links or arms. Little internal damping. Little or no maintenance. High travel
Leaf springs	These springs can be single or multiple leaf. They are most often mounted longitudinally. Nowadays they are only ever used on commercial vehicles	Can transmit longitudinal and lateral forces. Short travel. High internal damping. High load capacity. Maintenance may be required. Low height but high weight.
Torsion bar springs (Figure 7.20)	A torsion bar is a spring where twisting loads are applied to a steel bar. They can be round or square section, solid or hollow. Their surface must be finished accurately to eliminate pressure points, which may cause cracking and fatigue failure. They can be fitted longitudinally or laterally	Maintenance free but can be adjusted. Transmit longitudinal and lateral forces. Limited self damping. Linear rate. Low weight. May have limited fatigue life
Rubber springs (Figure 7.21)	Nowadays rubber springs are only used as a supplement to other forms of springs. They are, however, popular on trailers or caravans	Progressive rate. Transmit longitudinal and lateral forces. Short travel. Low weight and low cost. Their springing and damping properties can change with changes in temperature

Figure 7.19 Coil spring

1. Support bracket
2. Actuating lever pivot bush
3. Guide tube
4. Locking fixture
5. Torsion bar
6. Internal guide bush
7. Actuating lever with guide bush

Figure 7.20 Torsion bar in guide tube

1. Hollow rubber spring
2. Rubber torsion spring
3. Rubber
4. Housing

Figure 7.21 Rubber springs

Name	Comments	Characteristics
Air springs (Figure 7.22)	Air springs can be thought of as being like a balloon or football on which the car is supported. The system involves compressors and air tanks. They are not normally used on light vehicles	Expensive. Good quality ride. Electronic control can be used. Progressive spring rate. High production cost

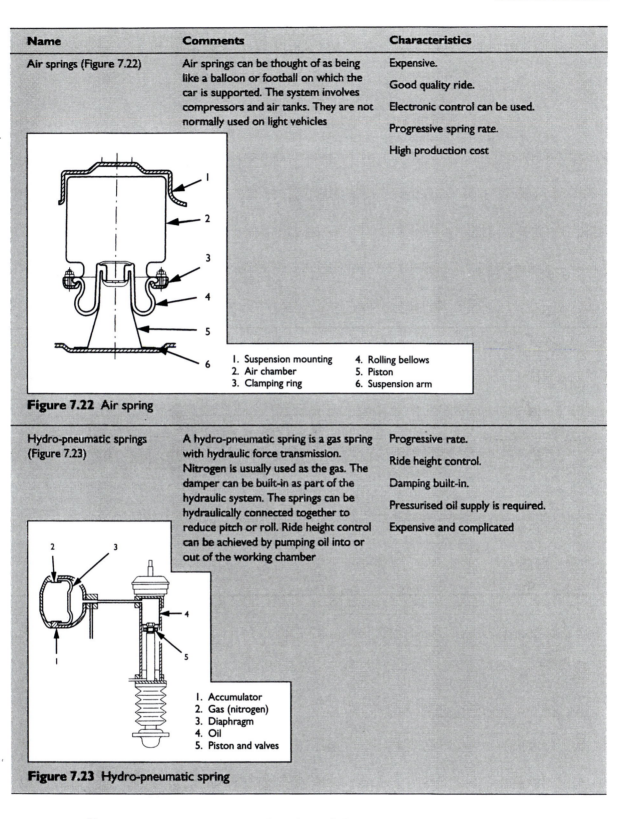

1. Suspension mounting
2. Air chamber
3. Clamping ring
4. Rolling bellows
5. Piston
6. Suspension arm

Figure 7.22 Air spring

Hydro-pneumatic springs (Figure 7.23)	A hydro-pneumatic spring is a gas spring with hydraulic force transmission. Nitrogen is usually used as the gas. The damper can be built-in as part of the hydraulic system. The springs can be hydraulically connected together to reduce pitch or roll. Ride height control can be achieved by pumping oil into or out of the working chamber	Progressive rate. Ride height control. Damping built-in. Pressurised oil supply is required. Expensive and complicated

1. Accumulator
2. Gas (nitrogen)
3. Diaphragm
4. Oil
5. Piston and valves

Figure 7.23 Hydro-pneumatic spring

Dampers Have you remembered to call the suspension component that reduces oscillation a damper, and not a shock absorber? Well done! The functions of a damper can be summarised as follows:

- ensure directional stability
- ensure good contact between the tyres and the road
- prevent build up of vertical movements
- reduce oscillations
- reduce wear on tyres and chassis components

There are a number of different types of damper:

Friction damper	Not used on cars today but you will find this system used as part of caravan or trailer stabilisers
Lever type damper	Used on earlier vehicles, the lever operates a piston which forces oil into a chamber
Twin tube telescopic damper (Figure 7.24)	This is the most commonly used type of damper. It consists of two tubes: an outer tube forming a reservoir space and containing the oil displaced from an inner tube. Oil is forced through a valve by the action of a piston as the damper moves up or down. The reservoir space is essential to make up for the changes in volume as the piston rod moves in and out

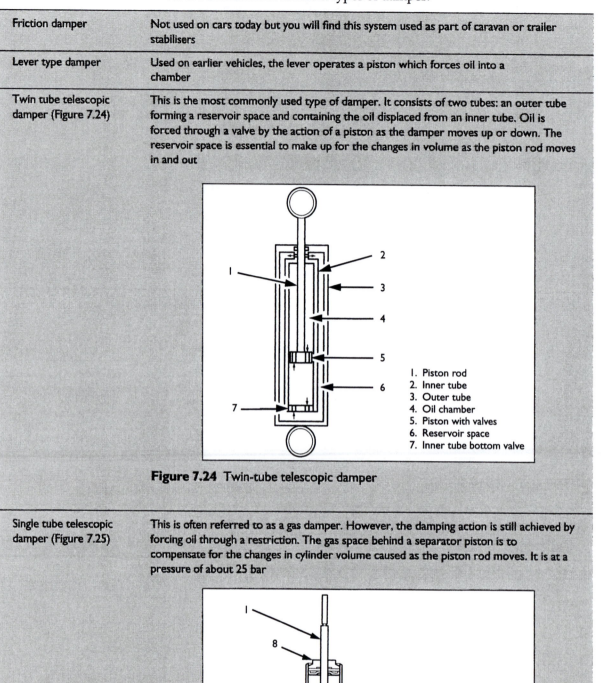

1. Piston rod
2. Inner tube
3. Outer tube
4. Oil chamber
5. Piston with valves
6. Reservoir space
7. Inner tube bottom valve

Figure 7.24 Twin-tube telescopic damper

Single tube telescopic damper (Figure 7.25)	This is often referred to as a gas damper. However, the damping action is still achieved by forcing oil through a restriction. The gas space behind a separator piston is to compensate for the changes in cylinder volume caused as the piston rod moves. It is at a pressure of about 25 bar

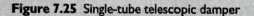

1. Piston rod
2. Working piston
3. Tube
4. Bush
5. Gas chamber
6. Separator piston
7. Oil chamber
8. Seal

Figure 7.25 Single-tube telescopic damper

Twin tube gas damper (Figure 7.26)

The twin tube gas damper is an improvement on the well-used twin-tube system. The gas cushion is used in this case to prevent oil foaming. The gas pressure on the oil prevents foaming which, in turn, ensures constant operation under all operating conditions. Gas pressure is lower than for a single tube damper at about 5 bar

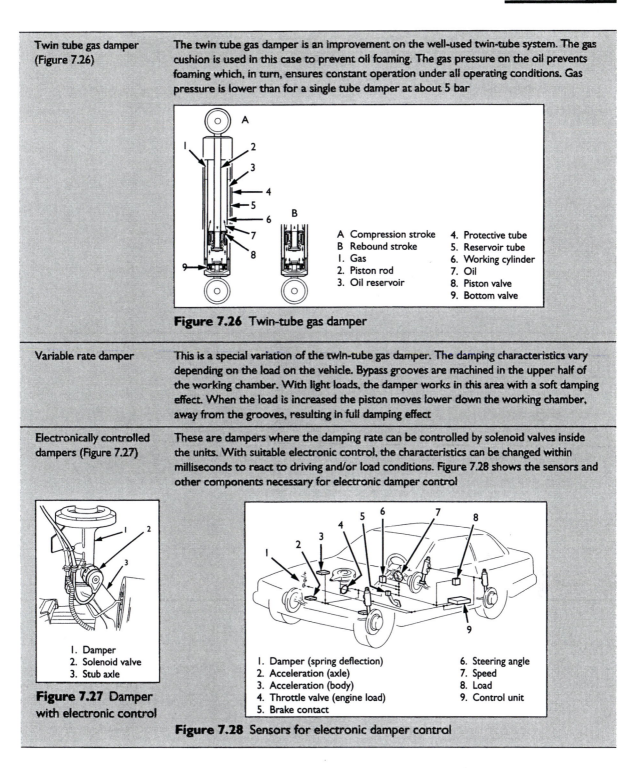

A Compression stroke	4. Protective tube
B Rebound stroke	5. Reservoir tube
1. Gas	6. Working cylinder
2. Piston rod	7. Oil
3. Oil reservoir	8. Piston valve
	9. Bottom valve

Figure 7.26 Twin-tube gas damper

Variable rate damper

This is a special variation of the twin-tube gas damper. The damping characteristics vary depending on the load on the vehicle. Bypass grooves are machined in the upper half of the working chamber. With light loads, the damper works in this area with a soft damping effect. When the load is increased the piston moves lower down the working chamber, away from the grooves, resulting in full damping effect

Electronically controlled dampers (Figure 7.27)

These are dampers where the damping rate can be controlled by solenoid valves inside the units. With suitable electronic control, the characteristics can be changed within milliseconds to react to driving and/or load conditions. Figure 7.28 shows the sensors and other components necessary for electronic damper control

1. Damper
2. Solenoid valve
3. Stub axle

Figure 7.27 Damper with electronic control

1. Damper (spring deflection)	6. Steering angle
2. Acceleration (axle)	7. Speed
3. Acceleration (body)	8. Load
4. Throttle valve (engine load)	9. Control unit
5. Brake contact	

Figure 7.28 Sensors for electronic damper control

LEARNING TASKS

➡ Look back at the key words. Explain each one to a friend, and/or write out a short description to keep as evidence.

➡ Make a simple sketch to show a McPherson strut layout.

➡ Examine several real systems and note the type of suspension system in use.

➡ Using other reference sources, write a short explanation about how to test a damper.

3 Suspension case study – Ford Scorpio

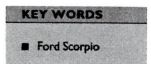
The Ford Scorpio has independent front suspension (Figure 7.29). McPherson struts and forged transverse arms are linked with the anti-roll bar on each side. As well as reducing body roll when cornering, the anti-roll bar locates the transverse arms and, as such, transmits braking, steering and driving forces. The anti-roll bar is attached to the body by special resilient bushes. A

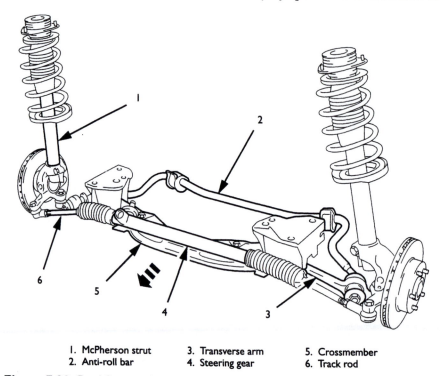

1. McPherson strut	3. Transverse arm	5. Crossmember
2. Anti-roll bar	4. Steering gear	6. Track rod

Figure 7.29 Ford Scorpio front suspension

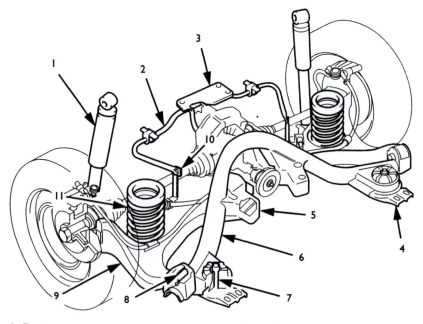

1. Damper	7. Rubber mount for crossmember
2. Anti-roll bar	8. Semi-trailing arm outer rubber mount
3. Differential housing mounting bracket	9. Semi-trailing arm
4. Crossmember locating plate	10. Anti-roll bar link rod
5. Semi-trailing arm inner rubber mount	11. Coil spring
6. Crossmember	

Figure 7.30 Ford Scorpio rear suspension

small amount of free play under light loads improves ride comfort. When load increases, a harder section of the bushes is reached. Camber and castor cannot be adjusted as they are determined by the chassis and body components.

The rear suspension is of the semi-trailing type with double-jointed drive shafts (Figure 7.30). Coil springs provide the spring medium together with double-acting dampers. The rear axle cross member is mounted on rubber bushes and attached to the body by locating plates. The final drive assembly is mounted to the body as this reduces unsprung mass. Anti-roll bars are used which are attached to the semi-trailing arms by link rods.

LEARNING TASK

➡ Examine real systems and compare their relative advantages and disadvantages.

8 Steering

1 Introduction

Start here! Most light vehicle are steered by the front wheels; the rear wheels then follow the front but on a smaller radius. Two factors which affect the steering effort must be taken into account when designing a steering system:

■ axle loading
■ contact area of the steered wheels.

Steering effort can be quite considerable, particularly with vehicles fitted with wide tyres. Power steering reduces this effort and increases safety and comfort. Typical capabilities of a steering system can be noted as follows:

■ the wheels must return to the central position after cornering – called the self-centring effect
■ two rotations of the steering wheel will produce a steering angle of about 40°.

Fitting steering dampers can reduce the influence of uneven road surfaces, but these are rare on normal road vehicles. It is important to remember the interaction between the steering system and the suspension system. Steerability of the vehicle depends very much on wheel adhesion, which in turn depends very much on the suspension system. Figure 8.1 shows the interaction between steering and suspension.

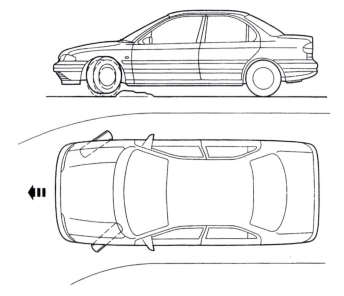

Figure 8.1 Interaction of steering and suspension

Terminology

Steering geometry	The angles at which the wheels are set relative to each other, the vehicle and the point about which the vehicle is turning. Figure 8.2 shows the main front axle steering geometry

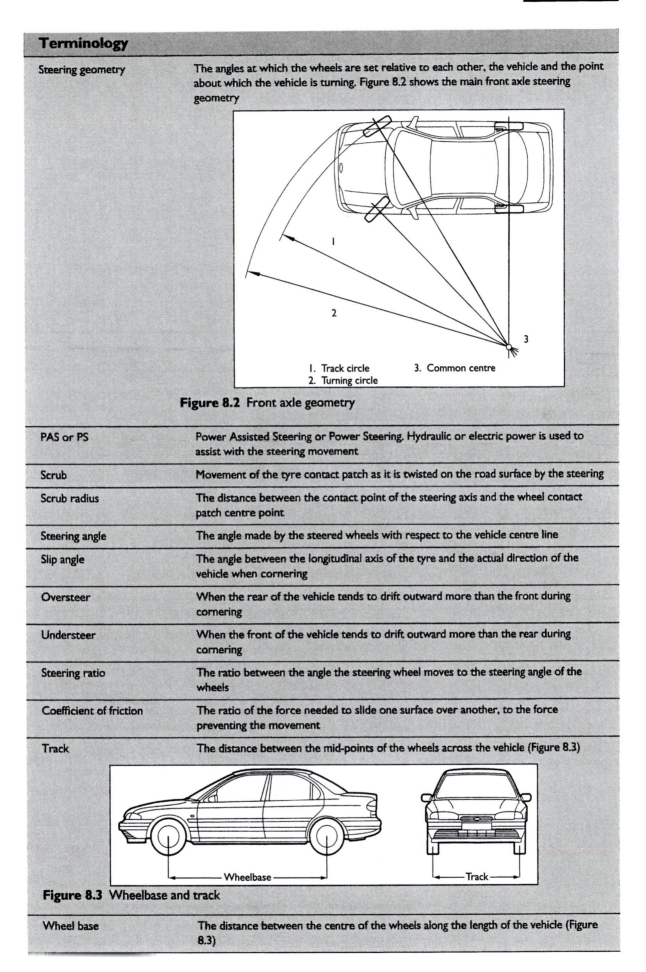

Figure 8.2 Front axle geometry

PAS or PS	Power Assisted Steering or Power Steering. Hydraulic or electric power is used to assist with the steering movement
Scrub	Movement of the tyre contact patch as it is twisted on the road surface by the steering
Scrub radius	The distance between the contact point of the steering axis and the wheel contact patch centre point
Steering angle	The angle made by the steered wheels with respect to the vehicle centre line
Slip angle	The angle between the longitudinal axis of the tyre and the actual direction of the vehicle when cornering
Oversteer	When the rear of the vehicle tends to drift outward more than the front during cornering
Understeer	When the front of the vehicle tends to drift outward more than the rear during cornering
Steering ratio	The ratio between the angle the steering wheel moves to the steering angle of the wheels
Coefficient of friction	The ratio of the force needed to slide one surface over another, to the force preventing the movement
Track	The distance between the mid-points of the wheels across the vehicle (Figure 8.3)

Figure 8.3 Wheelbase and track

Wheel base	The distance between the centre of the wheels along the length of the vehicle (Figure 8.3)

Steering characteristics

The steering characteristics of a vehicle, or in other words the way in which it reacts when cornering, can be described by one of three headings:

- oversteer (Figure 8.4)
- understeer (Figure 8.5)
- neutral (Figure 8.6).

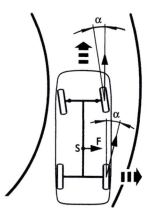

S Centre of gravity behind the vehicle centre
F Centrifugal force
α Slip angle

Figure 8.4 Oversteer

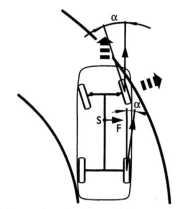

S Centre of gravity in front of the vehicle centre
F Centrifugal force
α Slip angle

Figure 8.5 Understeer

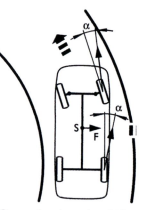

S Centre of gravity at the vehicle c
F Centrifugal force
α Slip angle

Figure 8.6 Neutral steer

These situations arise because of the way the force acting on the vehicle's centre of gravity is distributed between the front and rear wheels. The centre of gravity position depends on the general layout of the vehicle.

Oversteer occurs when the rear of the vehicle tends to swing outward more than the front during cornering. This is because the slip angle on the rear axle is significantly greater than the front axle. This causes the vehicle to travel in a tighter circle, hence the term oversteer. If the steering angle is not reduced, the vehicle will break away and all control will be lost. Turning the steering towards the opposite lock will reduce the front slip angle.

Understeer occurs when the front of the vehicle tends to swing outward more than the rear during cornering. This is because the slip angle on the rear axle is significantly smaller than the front axle. This causes the vehicle to travel in a greater circle, hence the term understeer. If the steering angle is not increased, the vehicle will be carried out of the corner and all control will be lost. Turning the steering further into the bend will increase the front slip angle. Front-engined vehicles tend to understeer because the centre of gravity is situated in front of the vehicle centre. The outward centrifugal force therefore has a greater effect on the front wheels than on the rear.

Neutral steering occurs when the centre of gravity is at the vehicle centre and the front and rear slip angles are equal. The cornering forces are therefore uniformly spread. Note, however, that understeer or oversteer can still occur if the cornering conditions change.

LEARNING TASKS

➡ Look back at the key words. Explain each one to a friend, and/or write out a short description to keep as evidence.

➡ Make a simple sketch to show the main steering angles – try not to just copy my pictures!

➡ Examine a real system and note the way components are connected to ensure the steering geometry is set.

2 Steering geometry

Camber ■ Typical value is about 0.5° (values will vary so check specs.)

On many cars, the front wheels are not mounted vertically to the road surface. Often they are tilted outwards at the top. This is called positive camber (Figure 8.7) and has the following effects:

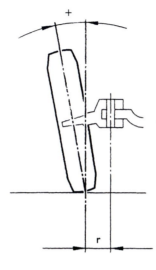

+ Positive camber angle
r   Scrub radius

Figure 8.7 Positive camber

■ easier steering, less turning effort required
■ less wear on the steering linkages
■ less stress on main components
■ smaller scrub radius, which reduces the effect of wheel forces on the steering.

Negative camber (Figure 8.8) has the following effect:

■ good cornering force.

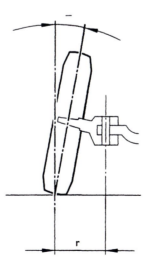

− Negative camber angle
r   Scrub radius

Figure 8.8 Negative camber

Some cars have rear wheels with negative camber. With independent suspension systems, wheels can change their camber from positive, through neutral, to negative as the suspension is compressed. This varies, though, with the design and position of the suspension hinge points.

Castor
- Typical value is about 2° to 4° (values will vary so check specs.)

The front wheels tend to straighten themselves out after cornering. This is due to a castor action. Supermarket trolley wheels automatically run straight when pushed because the axle on which they rotate is behind the swivel mounting. Vehicle wheels get the same result by leaning the swivel pin mountings back so that the wheel axle is moved slightly behind the line of the swivel axis. The further the axle is behind the swivel, the stronger the straightening effect will be. The main reasons for a positive castor angle (Figure 8.9) are:

- self centring action
- helps to determine the steering torque when cornering.

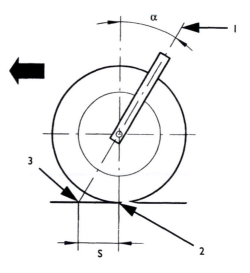

1. Steering axis
2. Wheel contact point
3. Positive castor, the point of intersection of steering axis with road surface
S Castor trail
α Castor angle

Figure 8.9 Positive castor

Negative castor (Figure 8.10) is used on some front wheel drive vehicles to reduce the return forces when cornering. Note a combination of steering geometry angles is used to achieve the desired effect. This means that, in some cases, the swivel axis produces the desired self centre action, so the castor angle may need to be negative in order to reduce the return forces on corners.

Swivel axis inclination
- Typical value is about 7° to 9° (values will vary so check specs.)

The swivel axis is also known as the steering axis. Swivel axis inclination (Figure 8.11) means the angle compared to vertical made by the two swivel joints when viewed from the front or rear. On a strut type suspension system the angle is broadly that made by the strut. This angle always leans in towards the middle of the vehicle. The swivel axis inclination (also called kingpin inclination) is mainly for:

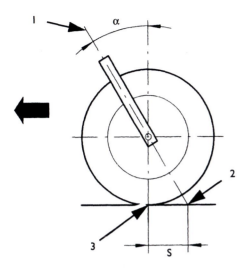

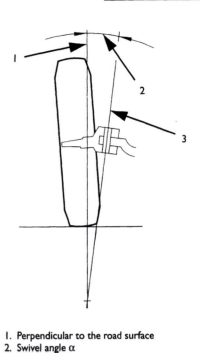

1. Steering axis
2. Negative castor, the point of intersection of
   steering axis with road surface
3. Wheel contact point
S Negative castor trail
α Negative castor angle (negative castor)

Figure 8.10 Negative castor

1. Perpendicular to the road surface
2. Swivel angle α
3. Steering axis

Figure 8.11 Swivel axis inclination

- producing a self centre action
- improving steering control on corners
- lighter steering action.

Scrub radius, wheel camber and swivel axis inclination all have an effect on one another. The swivel axis inclination mainly affects the self-centring action, also known as the aligning torque. Because of the axis inclination, the vehicle is raised slightly at the front as the wheels are turned. Therefore, the weight of the vehicle tries to force the wheels back into the straight-ahead position.

Let's look at this another way. If you look at a vehicle on a wheel free lift, you would notice that, as the steering is turned, the front wheels describe an arc determined by the scrub radius. Because of the swivel axis inclination this arc is on a sloping plane (Figure 8.12). The result of this is that, when on a solid road, the wheels cannot move down the plane so the front of the

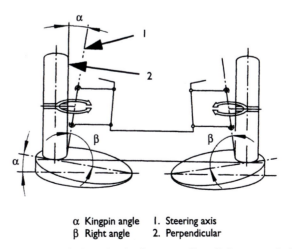

α Kingpin angle   1. Steering axis
β Right angle   2. Perpendicular

Figure 8.12 When turned, the wheels theoretically roll downwards, but in practice they raise the front of the car

vehicle lifts up. As a result, the wheels will always try to return to the straight-ahead position. This is a particular advantage on rough surfaces as the restoring force of the swivel axis inclination counteracts the disturbing forces.

Tracking   As a front wheel drive car drives forward, the tyres pull on the road surface, taking up the small amount of free play in the mountings and joints. For this reason the tracking is often set toe-out so the result is that the wheels point straight ahead when the vehicle is moving. Rear wheel drive tends to make the opposite happen because it pushes against the front wheels. The front wheels are therefore set toe-in. When the car moves, the front wheels are pushed out, taking up the slack in the joints, so the wheels again end up straight-ahead. The amount of toe-in or toe-out is very small, normally not exceeding 5 mm (the difference in the distance between the front and rear of the front wheels). Correctly set tracking ensures: true rolling of the wheels (therefore reducing tyre wear).

Figure 8.13 shows wheels set toe-in and toe-out.

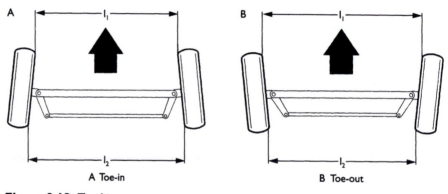

Figure 8.13 Tracking

Scrub radius   The scrub radius is the distance between the contact point of the steering axis with the road and the wheel centre contact point. The purpose of designing-in a scrub radius is to reduce the steering force and to prevent steering shimmy. It also helps to stabilise the straight-ahead position.

It is possible to design the steering with a negative, positive or zero scrub radius.

KEY WORDS

- All words in the table
- Toe-out on turns

Scrub radius	Description	Properties
Negative (Figure 8.14) 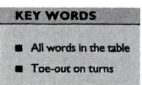	The contact point of the steering axis hits the road between the wheel centre and the outer edge of the wheel	Braking forces produce a torque which tends to make the wheel turn inwards. The result of this is that the wheel with the greatest braking force is turned in with greater torque. This steers the vehicle away from the side with the heaviest braking producing a built-in counter steer action which has a stabilising effect

Figure 8.14 Negative scrub radius

Scrub radius	Description	Properties
Positive (Figure 8.15) 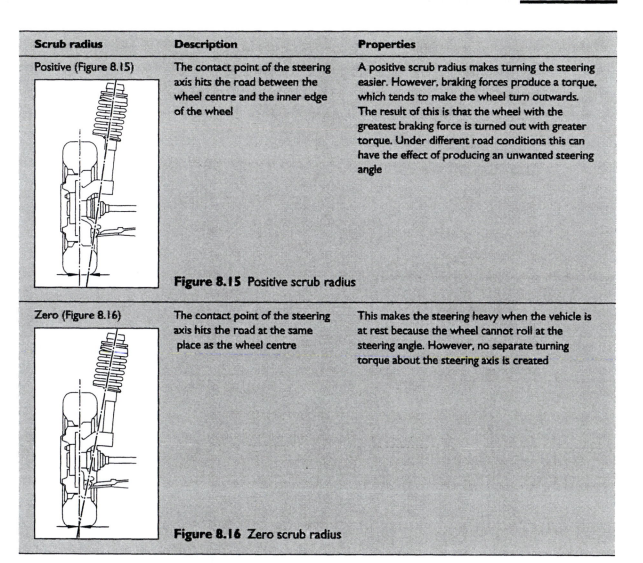 **Figure 8.15** Positive scrub radius	The contact point of the steering axis hits the road between the wheel centre and the inner edge of the wheel	A positive scrub radius makes turning the steering easier. However, braking forces produce a torque, which tends to make the wheel turn outwards. The result of this is that the wheel with the greatest braking force is turned out with greater torque. Under different road conditions this can have the effect of producing an unwanted steering angle
Zero (Figure 8.16) **Figure 8.16** Zero scrub radius	The contact point of the steering axis hits the road at the same place as the wheel centre	This makes the steering heavy when the vehicle is at rest because the wheel cannot roll at the steering angle. However, no separate turning torque about the steering axis is created

From the information given, you will realise that the decisions about steering geometry are not clear-cut. One change may have a particular advantage in one area but a disadvantage in another.

Ackerman principal

Not surprisingly, the Ackerman principal or the Ackerman steering system was named after Dr. Ackerman! All 'normal use' motor vehicles use this system, which is also known as stub axle steering. The principal is that, as a vehicle turns a corner, the inner wheel follows a sharper curve than the outer. To ensure true rolling, the inner wheel must turn through a sharper angle. Remember that true rolling of a steered wheel is when a line at 90° from each wheel passes through the imaginary common centre (Figure 8.17).

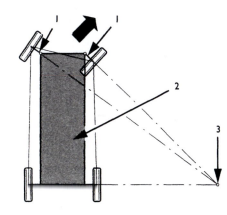

Figure 8.17 Ackerman steering

1. Stub axle
2. Area of support
3. Imaginary common centre

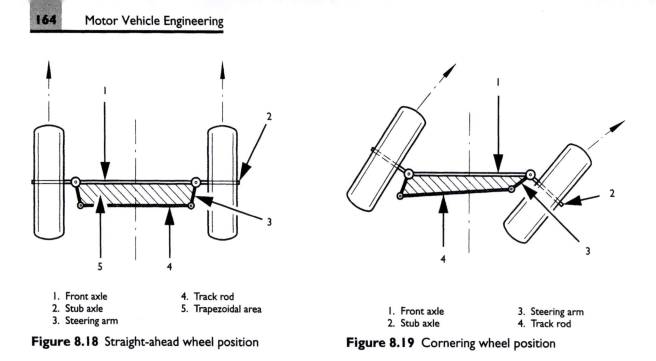

I. Front axle
2. Stub axle
3. Steering arm

4. Track rod
5. Trapezoidal area

Figure 8.18 Straight-ahead wheel position

I. Front axle
2. Stub axle

3. Steering arm
4. Track rod

Figure 8.19 Cornering wheel position

To achieve this, the track rod (usually the rack) and the steering arms are made in the form of a trapezium. When the vehicle is moving straight ahead, the track rod and the front axle are parallel to one another (Figure 8.18). When cornering, the stub axles are swivelled to turn the front wheels. As the wheels turn, the track rod is made to lie at an increasing angle to the axle (Figure 8.19). The result is that the two steering arms experience different amounts of travel and hence the inside wheel is turned more than the outer.

Because the tracking of the wheels changes as the steering is turned, the difference in angle is often referred to as 'toe-out on turns'.

LEARNING TASKS

➡ Look back at the key words. Explain each one to a friend, and/or write out a short description to keep as evidence.

➡ Make a simple model using card and drawing pins to demonstrate the Ackerman steering principal.

➡ Examine a real system and note the layout of steering components to achieve true rolling on corners.

3   Steering systems

Rack and pinion   The steering rack (Figure 8.20) is now used almost without exception on light vehicles. This is because it is simple in design and very long lasting. The wheels turn on two large swivel joints. Another ball joint (often called a track rod end) is fitted on each swivel arm. A further ball joint to the ends of the rack connects the track rods. The rack is inside a lubricated tube and gaiters protect the inner ball joints. The pinion meshes with the teeth of the rack and, as it is turned by the steering wheel, the rack is made to move back and forth, turning the front wheels on their swivel ball joints.

On many vehicles now, the steering rack is still used but is augmented with hydraulic power assistance.

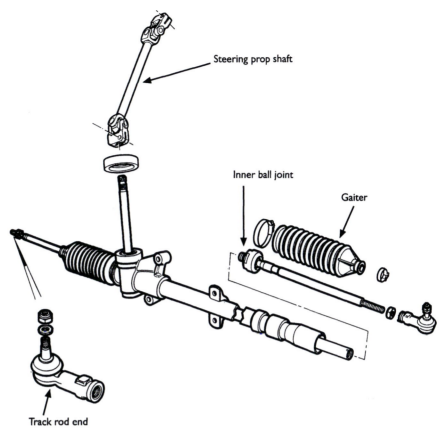

Steering prop shaft

Inner ball joint

Gaiter

Track rod end

Figure 8.20 Steering rack

Hydraulic power steering

To steer the front wheels, a certain effort is required. This is primarily determined by axle load. The effort required is most noticeable under the following conditions:

■ low speed
■ low tyre pressures
■ larger tyre contact patch
■ tight cornering.

Generally the force required to turn the steering wheel should not exceed 250 N, hence the need for some assistance. Power steering or power assisted steering, as it is often known, has been developed to meet the following requirements:

■ fail safe – in other words, steering must still be possible if a power assistance fault develops
■ precise onset of power assistance.

Power steering systems fall into one of two categories:

1 Modular design where the hydraulic ram is part of the rack (Figure 8.21).
2 Semi-modular design where the hydraulic ram is separate from the rack (Figure 8.22).

Figure 8.23 shows a typical power-assisted steering system. Similar designs are used on a wide variety of light vehicles.

Hydraulic fluid is pressurised by an engine driven pump, which takes its supply from a reservoir. When the steering wheel is turned, the control valve allows a set amount of fluid pressure to the appropriate side of the ram

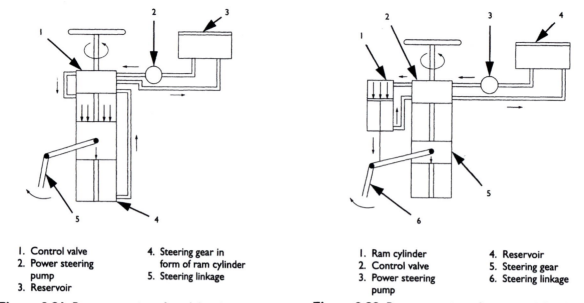

1. Control valve
2. Power steering pump
3. Reservoir
4. Steering gear in form of ram cylinder
5. Steering linkage

Figure 8.21 Power steering of modular design

1. Ram cylinder
2. Control valve
3. Power steering pump
4. Reservoir
5. Steering gear
6. Steering linkage

Figure 8.22 Power steering of semi-modular design

piston in the cylinder. It is this pressure which assists with the steering force. The amount of pressure is determined by the amount of force exerted on the control valve.

A spring element in the control valve determines how much power assistance is required. This element is usually a torsion bar that takes the turning force from the steering wheel. Its twist is limited by splines to just a few degrees, for safety reasons and to ensure the correct opening of the control valve. If you imagine a car that is not moving, and the steering wheel

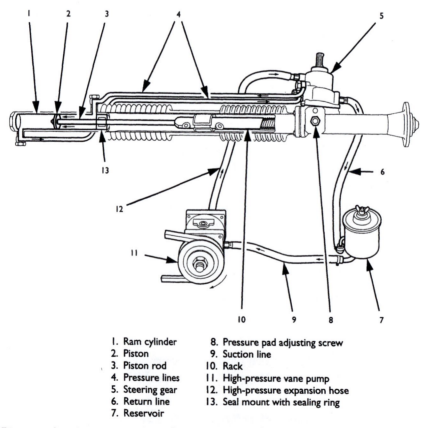

1. Ram cylinder
2. Piston
3. Piston rod
4. Pressure lines
5. Steering gear
6. Return line
7. Reservoir
8. Pressure pad adjusting screw
9. Suction line
10. Rack
11. High-pressure vane pump
12. High-pressure expansion hose
13. Seal mount with sealing ring

Figure 8.23 Diagram showing construction of a power-assisted steering system

is turned to try and move the wheels, the force required will twist the torsion bar to its full extent, one way or the other. This opens the control valve and supplies full fluid pressure to the appropriate side of the ram. On the other hand, when the vehicle is moving and only a small force is applied to the steering wheel, the torsion bar will only twist slightly thus allowing a small amount of power assistance.

Return fluid from the ram flows back to the reservoir. When the steering movement is interrupted, in other words, when no force is applied to the steering wheel, the control valve takes up a central position and no fluid is forced into the ram. If the steering wheel is now turned in one direction or another, the pressure will again be applied to the ram. Figure 8.24 shows the main parts of a typical control valve and its operation.

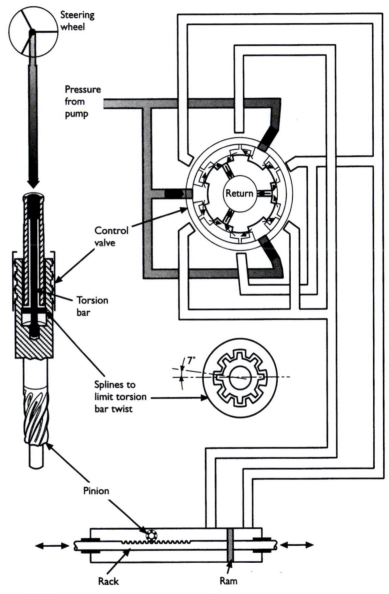

Figure 8.24 Power steering system operation

Electric power steering

There are three types of electric power steering currently at various stages of development (Vauxhall have a system in production). Respectively, they operate by:

- replacing the conventional hydraulic pump with an electric motor whilst the ram remains much the same

- a drive motor, directly assisting with the steering, which itself has no hydraulic components
- active steering, where the steering wheel is replaced with a joystick!

The first of these systems shows some scope for development, as the pump will only run when needed. This reduces fuel consumption and also allows the drive belt arrangement at the front of the engine to be simplified.

The second system listed looks the most promising. Lucas Advanced Engineering have developed an electronically controlled power steering system. A small electric motor acts directly on the steering via an epicyclic gear train. This completely replaces the hydraulic pump and servo cylinder. It also eliminates the fuel penalty of the conventional pump and greatly simplifies the drive arrangements. Engine stall when the power steering is operated at idle speed is also eliminated.

An optical torque sensor is used to measure driver effort on the steering wheel. The sensor works by measuring light from an LED which shines through holes that are aligned in discs at either end of a 50 mm torsion bar fitted into the steering column. This system occupies little under-bonnet space, something which is at a premium these days. The 400 W motor only averages about 2 amps current in urban driving conditions. The cost benefits over conventional hydraulic methods are expected to be considerable.

Steering towards safety and comfort

As you may recall from the level 2 book, safety of vehicle systems can be placed under two headings:

- active
- passive.

Active safety in the steering system relates to the feel of the steering and the general comfort of the driver. Driver feel is an important aspect of active safety. The steering should be light enough for comfort and ease of use but should also provide feedback of the road conditions. This is often described as the reversibility of the steering system.

Many manufacturers now build in adjustable steering positions for the driver's comfort. Figure 8.25 shows a tilt adjustable steering column. The double UJ and a sliding joint allow this movement. Figure 8.25 also shows a collapsible top mounting which is a passive safety feature. Figure 8.26

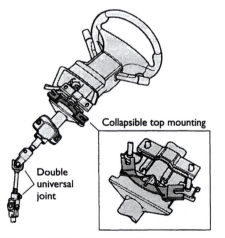

Figure 8.25 Tilt adjustable steering column

Figure 8.26 This steering column incorporates an energy absorbing structure that collapses in the event of a collision

illustrates an energy absorbing structure that collapses in the event of a collision. In conjunction with other safety features, this greatly reduces the risk of injury. Passive safety means that the system waits until you decide to have an accident!

Four-wheel steering   To understand four-wheel steering, it is first useful to recall or imagine the effects of rear wheel steering. If you have ever driven or watched the movement of a forklift truck, you will realise the different effect moving the rear wheels has on vehicle position. This is the same effect on a normal car when reversing – it is why some drivers have trouble reversing into a parking space or out of a garage! The key point is that the trailing end of the vehicle tends to slew in the direction that the wheels are turned, as shown in Figure 8.27.

When all four wheels are turned, the overall effect on the vehicle changes again. An interesting point, though, is that the effect varies (and I don't mean the direction the vehicle turns) depending on which way the rear wheels are moved, but the vehicle still moves the same direction. This is best illustrated by referring to Figure 8.28. The effects could be described as a turn or a drift.

Direction of front movement

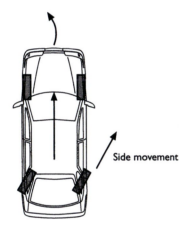

Side movement

Figure 8.27 Rear wheel steering

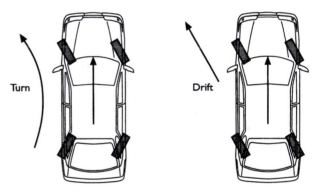

Turn

Drift

Figure 8.28 Four-wheel steering effects

Several manufacturers have developed four-wheel steer systems. However, it is not widely available as yet. This is because the advantages do not yet outweigh the considerable extra cost. Nonetheless, some systems are in use and do make use of the characteristics shown in Figure 8.28. At low speeds, the wheels are turned in opposite directions to improve the drag or slip on the tyres as well as reducing the turning circle. At high speeds the wheels are turned in the same direction such as for when changing lanes on a motorway. It is worth pointing out, though, that the amount of turn on the rear wheels is much less than the front.

Figure 8.29 shows the layout of the components on one system in current use. As is common with many if not all aspects of the vehicle, electronic control is now playing a role in four-wheel steering systems. This is used to determine the amount and direction of rear wheel movement.

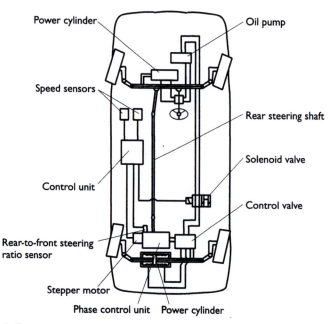

Figure 8.29 Four-wheel steering system

4 Power steering system – case study

Introduction It is not possible to cover every type of steering system on every vehicle in this book! However you should now be able to understand the principles and be able to find further information from other sources, as you need it. This short section will illustrate the system used on the Ford Probe GT. Figure 8.30 shows the system components.

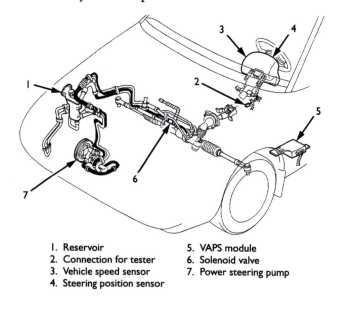

1. Reservoir
2. Connection for tester
3. Vehicle speed sensor
4. Steering position sensor
5. VAPS module
6. Solenoid valve
7. Power steering pump

Overview

Variable Assist Power Steering or VAPS is used on the GT version of the Ford Probe. Fluid pressure is controlled electronically. Variable power steering (sometimes called progressive power steering) makes steering easier at low speeds and provides good driver feel at higher speeds. The main components shown in Figure 8.30 are as follows:

■ steering position sensor (Figure 8.31)
■ VAPS electronic control unit
■ vehicle speed sensor
■ power steering solenoid valve.

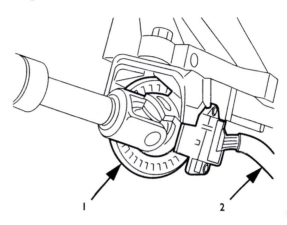

1. Pulse emitter disc
2. Steering position sensor

Figure 8.31 Steering position sensor

The electronic control unit monitors the signals from the vehicle speed sensor and the steering position sensor. From this data, it can then work out the power assistance required. The solenoid valve controls the amount of assistance, because the valve in turn controls fluid pressure. Maximum power assistance occurs at speeds less than 10 Km/h (6 mph) or when the steering wheel is rotated more than 45°.

Adjustment

The VAPS electronic control unit has an interesting feature. A slide switch on the side of the unit allows the following settings for the power steering:

■ switch position 'H': 10% greater effort required when steering
■ switch position 'N': normal adjustment
■ switch position 'L': 10% less effort required when steering.

LEARNING TASKS

➡ Look back at the key words. Explain each one to a friend, and/or write out a short description to keep as evidence.

➡ Make a list of the advantages or merits of the system described here.

5 Wheels and tyres

Introduction

All of the basic information about tyres is in the level 2 book. However, this section will describe further the operation and use of tyres on the vehicle. The information about wheel sizes and designation is included here for

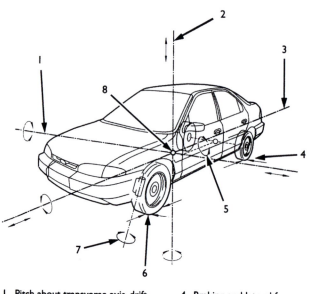

1. Pitch about transverse axis, drift along transverse axis
2. Yaw (side-slip) about vertical axis, bounce along vertical axis
3. Roll about longitudinal axis, jerking along longitudinal axis
4. Braking and lateral forces
5. Wheel tramp
6. Drive, braking and lateral forces
7. Shimmy about steering axis
8. Vehicle centre of gravity

Figure 8.32 Various forces acting on vehicle

completeness. The forces to be transmitted the road surface by the tyre is represented by Figure 8.32. By way of a reminder, the following list describes the functions of the tyres:

- support the weight of the vehicle
- ensure good road adhesion
- transmit driving, braking and cornering forces
- improve the ride comfort
- remain at the correct pressure and in good condition
- operate for a high mileage.

Wheel sizes Wheel sizes are often quite a mystery (Figure 8.33) and I would always recommend referring to manufacturer's data. However, a standard is used and is demonstrated in the following example. **7KX15H** is an example of a rim designation.

- 7 refers to the rim width in inches
- K refers to the shape and size of the rim flange (J and JK are also used)
- X when used, designates a drop centre or well based rim
- 15 is the rim diameter in inches
- H refers to an asymmetrical rim with a hump bead seat on the inner and outer sides (A is symmetrical and B is asymmetrical).

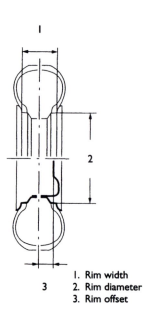

1. Rim width
2. Rim diameter
3. Rim offset

Figure 8.33 Wheel dimensions

Static coefficient of friction

When a force, such as the braking or driving force, is transmitted to the contact patch of a rolling wheel, a relative movement or slip occurs between the tyre and the road surface. This means that the distance covered by the vehicle is more or less than the rolling circumference of the tyre. This slip is expressed as a percentage and represents the difference between the distance covered by a rolling wheel with and without force applied. A locked wheel when braking would have a slip of 100%, for example.

The amount of slip varies depending on the following factors:

- drive or braking force
- cornering force
- static coefficient of friction between the tyre and the road surface.

The static coefficient of friction is determined by the road surface (e.g. concrete or asphalt), the condition of the road surface (e.g. dry or wet) and the characteristics of the tyre. The Greek letter µ – 'mu' – is used to denote the static coefficient of friction.

Side deflection and lateral forces

When a tyre is set at the correct pressure, a slight deformation of the tyre sidewall takes place in the region of the contact patch. As the vehicle rolls, the tyre also experiences circular deformation or flexing. This flexing produces a rolling resistance. The value of this resistance varies depending on:

- tyre cross section
- rubber compound
- tread pattern
- road surface
- vehicle speed.

A tyre is only able to transmit lateral forces when it is rolling at an angle to the direction of travel. This is why a tyre does not roll straight ahead when cornering but flexes laterally. Due to this flexing, the tyre develops a resistance or a side force that keeps the vehicle on course. The camber and toe-in of the wheels introduce the side deflection of the tyre.

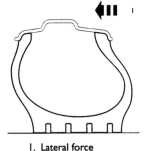

1. Lateral force

Figure 8.34 Tyres under lateral force loading

Apart from the centrifugal forces when cornering, it is also necessary to transmit lateral forces to absorb other disturbances such as side wind. Figure 8.34 shows a tyre under lateral force loading.

Slip angle

As discussed earlier, in relation to understeer and oversteer, at higher cornering speeds the centrifugal force drives the vehicle towards the outside of the curve. In order to keep the vehicle on track, the tyres must transmit cornering forces, which counteract the centrifugal force. This occurs as the tyre flexes laterally. Because of this, the wheels no longer move in their turned direction but drift at an angle to the direction of travel. This angle, which is the difference between the tyres longitudinal axis and the actual direction of vehicle movement when cornering, is known as the 'slip angle'. Figure 8.35 shows a tyre with the slip angle marked. The following two points controlled by the driver determine the value of the slip angle:

- vehicle speed
- steering angle.

Tyres corner best at a slip angle of between 15º and 20º. The value of the lateral adhesion, or how well the car 'sticks' to the road, is dependent on:

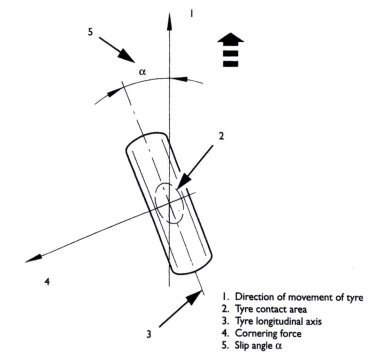

1. Direction of movement of tyre
2. Tyre contact area
3. Tyre longitudinal axis
4. Cornering force
5. Slip angle α

Figure 8.35 Slip angle

- slip angle
- wheel load
- static coefficient of friction.

In general, steering systems are designed to keep the front wheels almost parallel on bends with radii greater than 20 metres. On sharper bends (smaller radii), the inner wheel turns through a greater angle in accordance with the Ackerman principle.

Developments in vehicle tyres – Goodyear Infinitred case study

Goodyear have developed a tyre with a lifetime tread life limited warranty. The following description is from an advert for the tyre on the Goodyear Tyre & Rubber Co. web site. I will add no comments, as you should now be able to draw your own conclusions from manufacturer's data.

- Infinitred is the first all-season radial ever to be backed by a lifetime tread life limited warranty. If you wear them out in the first three years, we'll replace them free.
- Because Infinitred offers such an amazing tread life warranty, it also provides an incredible mile-per-dollar (pound?) economy. That means you get outstanding all-season performance and a lifetime tread life limited warranty, all at an unmatched value.
- It has deeper tread grooves than any of our other all-season passenger tyres for longer, more even wear and legendary Goodyear traction.
- An extra deep tread design and a premium compound deliver all-season performance throughout the life of the tyre.
- A wide tread and reinforced shoulder area deliver responsive handling, even wear and a quiet ride.
- Infinitred performs 19% better on rain-slick roads than a competitor's 80 000-mile tyre.

We gave Infinitred our deepest tread ever 12/32" (9.5 mm) to be exact in tyre size P175/70R13, plus deep blading for uniform wear and consistent performance throughout the life of the tyre. Inside, you'll find powerfully strong polyester cords, two durable high-tensile steel belts and high-tech polymer molecular compounds. It all adds up to a tyre so durable and dependable, it's covered by an impressive, unprecedented lifetime warranty, good for as long as you own your car.

Figure 8.36 shows a cutaway of the Infinitred tyre.

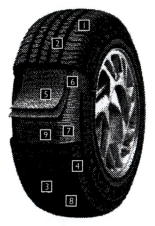

Figure 8.36 A new tyre design?

1. Premium all-weather tread compound for exceptional tread life, even wear and outstanding wet traction.
2. Extra-deep tread and angled tread blocks for long, even wear.
3. Deep design features with minimal blading for a tread design that does not change with wear.
4. Reinforced shoulder elements for responsive handling, even wear and excellent wet traction.
5. Unique encapsulating steel belt compound for added durability.
6. Two steel belts for strength and durability.
7. Radial ply construction and polyester cord body for a smooth, comfortable ride.
8. All-season tread designed for year-round traction.
9. Halobutyl liner for air retention over the life of the tyre.

LEARNING TASKS

➡ Look back at the key words. Explain each one to a friend, and/or write out a short description to keep as evidence.

➡ Consider the advantages and merits of an Infinitred tyre

➡ Examine a real system and note some tyre and wheel size markings.

9 Brakes

1 Introduction

Start here! The main purpose of the braking system is to slow or stop a vehicle. To do this, the vehicle's kinetic energy (movement energy) must be converted into another form of energy (heat). The friction between the brake pads and discs causes heat, taking energy away from the vehicle's movement. This is known as energy conversion (you will recall that energy cannot be destroyed, only converted to another form). The main braking system of a car works by hydraulics so, as the driver presses the brake pedal, liquid pressure forces pistons to apply brakes on each wheel.

Two types of brakes are used for light vehicles. Disc brakes are used on the front wheels of most cars. It is now common to find disc brakes on all wheels of sports and performance cars. In this system, braking pressure forces brake pads against both sides of a steel disc. Drum brakes are fitted on the rear wheels of some cars and on all wheels of older vehicles. Braking pressure forces brake shoes to expand outwards into contact with a drum. The important part of brake pads and shoes is a friction lining that grips well and withstands wear. Figure 9.1 shows a disc brake and Figure 9.2 a typical drum brake.

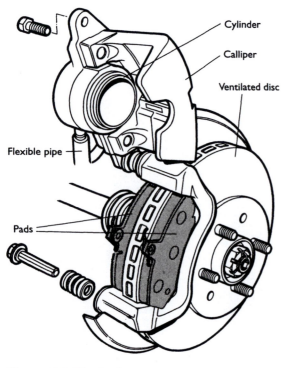

Figure 9.1 Disc brake

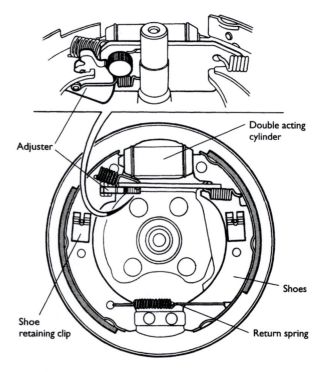

Figure 9.2 Drum brake

Terminology	
Split line	A safety system designed so that, if a serious problem such as a leak occurs in part of the braking system, at least half of the brakes will continue to work
Self servo action	The rotating action of a brake drum tends to pull the leading brake shoe harder into contact
Leading shoe	The brake shoe after the wheel cylinder in the direction of wheel rotation
Trailing shoe	The brake shoe before the wheel cylinder in the direction of wheel rotation
Load compensation	This system allows the braking pressure to the rear wheels to increase as load in the vehicle increases
Energy conversion	One form of energy being converted to another. In the braking system, the best example is the conversion of mechanical energy to heat energy at the discs, pads/drums and shoes
Brake fade	If the brakes become so hot that they cannot convert energy fast enough, the brakes will become much less efficient or, in other words, fade away! A form of brake fade can also be caused if the heat generated is enough to melt the bonding resin in the friction material. This reduces the frictional value of the linings or pads
ABS	Anti-lock brake system
Tandem master cylinder	A double acting cylinder used now on all vehicles as part of the split line system
Slave cylinder	Also called the wheel cylinder, it converts fluid pressure into movement to actuate the brake shoes or pads

KEY WORDS

- All words in the table

LEARNING TASKS

- Look back at the key words. Explain each one to a friend, and/or write out a short description to keep as evidence.
- Examine a real system and make a simple sketch to show the layout of the braking system components.
- Write down some advantages and disadvantages of ABS.

2 Braking system details

Split line systems   Safety is built into braking systems by using a double acting master cylinder (Figure 9.3). This is often described as 'tandem' and can be thought of as two master cylinders inside one housing. The pressure from the pedal acts on both cylinders but fluid cannot pass from one to the other. Each cylinder is then connected to a separate circuit. Figure 9.4 shows the most common, which is the diagonal split. Under normal operating conditions, the pressure developed in the first cylinder is transmitted to the second due to the fluid in the first chamber acting directly on the second piston.

Figure 9.5 shows how, if one line fails, the first piston meets no restriction and closes up to the second piston. Further movement will now provide pressure for the second circuit. The driver will notice that pedal travel increases, but some braking performance will remain. If the fluid leak is from the second circuit, then the second piston will meet no restriction and close up the gap. Braking will now be just from the first circuit.

Diagonal split brakes are the most common and are used on vehicles with a negative scrub radius. Steering control is maintained under brake failure conditions.

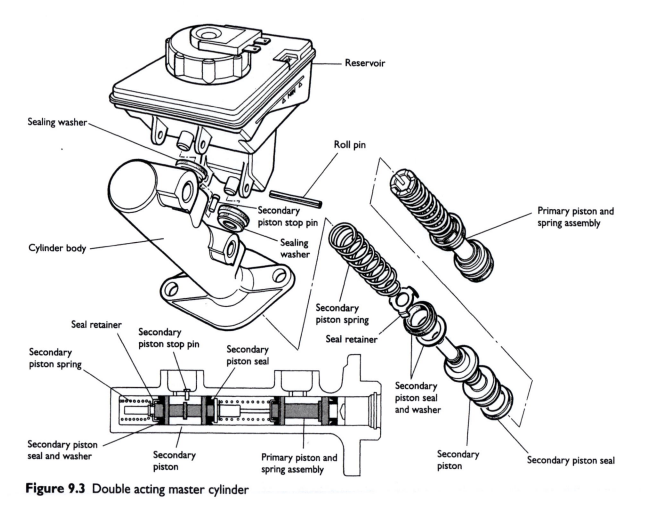

Figure 9.3 Double acting master cylinder

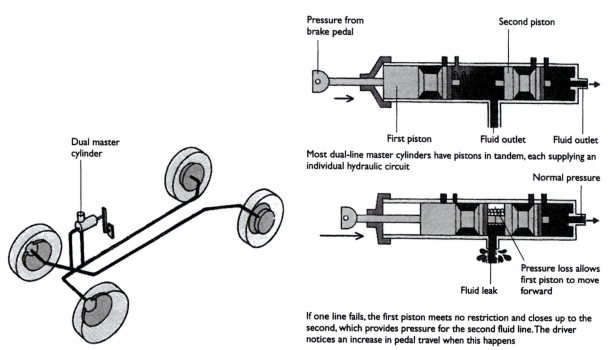

Pressure from brake pedal

Second piston

First piston   Fluid outlet   Fluid outlet

Most dual-line master cylinders have pistons in tandem, each supplying an individual hydraulic circuit

Normal pressure

Pressure loss allows first piston to move forward

Fluid leak

If one line fails, the first piston meets no restriction and closes up to the second, which provides pressure for the second fluid line. The driver notices an increase in pedal travel when this happens

Dual master cylinder

Figure 9.4 Diagonal split line brakes

Figure 9.5 Operation of a tandem master cylinder

Leading and trailing brake shoes

Brake shoes are mounted inside a cast iron drum. They are mounted on a steel backplate, which is rigidly fixed to a stationary part of the axle. The two curved shoes have friction material on their outer faces. One end of each shoe bears on a pivot point. The other end of each shoe is pushed out by the action of the wheel cylinder when the brake pedal is pressed. This puts the brake linings in contact with the drum inner surface. When the brake pedal is released, the return spring pulls the shoes back to their rest position.

The precise way in which the shoes move into contact with the drum affects the power of the brakes. If the shoes are both hinged at the same point then the system is said to have one leading and one trailing shoe. As the shoes are pushed into contact with the drum, the leading shoe is dragged by the drum rotation harder into contact, whereas the rotation tends to push trailing shoe away. This 'self-servo' action on the leading shoe can be used to increase the power of drum brakes. This is required on the front wheels of all-round drum brake vehicles. The shoes are arranged so that they both experience the self-servo action. The shoes are pivoted at opposite points on the backplate and two wheel cylinders are used. The arrangement is known as 'twin leading shoe brakes'. It is not suitable for use on the rear brakes because, if the car is travelling in reverse, it would become a twin trailing shoe arrangement, which means the efficiency of the brakes would be seriously reduced. Therefore the leading and trailing layout is used on rear brakes as one shoe will always be leading regardless of the direction the vehicle is moving in. Figure 9.6 shows the operation of leading and trailing shoes and twin leading shoes. The standard layout of drum brake systems is normally:

- twin leading shoe brakes on the front wheels
- leading and trailing shoe brakes on the rear wheels.

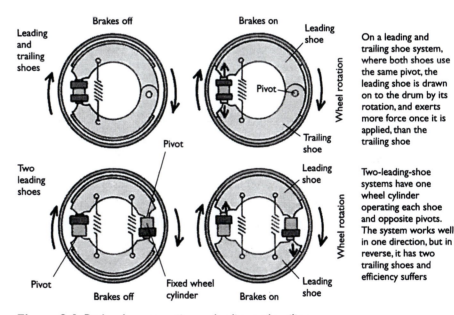

Figure 9.6 Brake shoe operation as leading and trailing

Disc brakes are now used on the front wheels of all light vehicles but many retain leading and trailing shoe brakes on the rear. In most cases, it is easier to attach a handbrake linkage to systems with shoes on the rear. This method will also provide the braking performance required when the vehicle is reversing.

Load compensation   The load on a vehicle determines the amount of braking which can be applied to the rear wheels before they lock up. As the load can vary between unladen to fully laden, the braking pressure which can be applied should also be varied. A load conscious pressure-reducing valve senses the weight on the rear wheels and adjusts the braking pressure accordingly. This still occurs even if the alteration occurs by weight transfer during braking. The

KEY WORDS

■ Self-servo action

■ Leading

■ Trailing

■ Tandem master
cylinder

■ Split line system

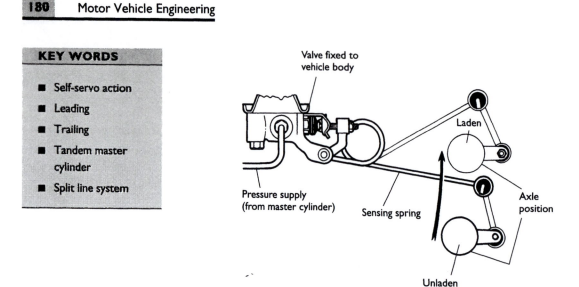

Figure 9.7 Load compensation valve sensing spring movement

valve is fitted in the rear line and has no effect on the front brakes. The unit is fitted to the underside of the vehicle body, close to the rear axle and the spring is connected to a point on the axle (Figure 9.7).

At normal braking, there is equal pressure all round and fluid pressure from the master cylinder passes through the open valve to the rear brakes. As the pressure builds up in the rear brake system, it acts on the valve until the force of the springs 'A' is overcome. The pistons now move to the right onto the rubber seal, closing the valve and sealing off further fluid pressure to the rear brakes. If braking is increased after the initial application, the valve allows only a proportion of any increase in fluid pressure to reach the rear brakes. There will now be considerably less pressure in the rear cylinders than in the front.

At all points between laden and unladen, the sensing spring applies force to the pistons in proportion to the weight being carried. Fluid pressure to the brakes has to be proportionally higher before the piston can be moved to the right to close the valve. Figure 9.8 shows a cross section of a compensation valve.

The load applied by the sensing spring to the piston will either:

■ increase and move the piston up, allowing increased pressure to the rear brakes

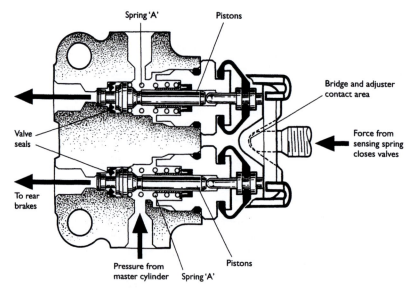

Figure 9.8 Load compensation unit (valves open)

- decrease and allow the piston to move down, thus reducing the pressure in the rear line.

The signal is instantly corrected in this way the whole time the brakes are applied. The fluid pressure to the rear brakes for any particular condition depends on the characteristics of the valve and the sensing spring rate, which are tailored to suit the vehicle for which the valve is designed.

The control of fluid pressure, by the valves, centres on a special seal. This has ribs spaced around its top face and part way down its edge. On the underside of the seal are a number of raised dimples. With only light pressure being applied to the brakes, the valve is open and inoperative. The spring pressure holds the valve stem against the dimples, which in turn allow the pressure to pass between them to the rear brakes. When the cut-in pressure is reached, the pressure being applied to the rear brakes causes the piston to move to the right and the underside of the piston head seals against the radius on the inner diameter of the seal.

An increase in master cylinder pressure produces an increase in input pressure in the unit. This, in turn, together with the spring pressure, causes the piston to move to the left and open the valve. This allows a further pressure increase to the rear brakes. It is important to note that, due to the piston design, once the cut-in pressure is reached, any further increase in master cylinder pressure will result in only a proportion of that pressure being passed to the rear brakes. When the master cylinder pressure is released, the outer lips of the seal collapse inward and allow fluid to pass back towards the master cylinder.

Antilock brake systems

Driving along a section of slippery road soaked by rain, you suddenly stamp as hard as you can on the brake pedal. In an 'ordinary' car, the brakes lock, the steering wheel becomes useless and the car skids out of control. A vehicle with ABS, however, stops straighter and better. Do remember though that ABS is not magic. Stopping on sheet ice, for example, is still very difficult, simply because of the lack of friction. Further details relating to ABS operation are included in chapter 11.

LEARNING TASKS

➡ Look back at the key words. Explain each one to a friend, and/or write out a short description to keep as evidence.

➡ Make a simple sketch to show a brake circuit including a load compensation valve.

➡ Write a short explanation about why twin leading shoe brakes are not used on the rear of a vehicle.

---

3   Power assisted brakes

Introduction

The brakes of a vehicle must perform well, whilst the effort required by the driver is kept to a reasonable level. This is the task of the brake servo. It is sometimes called a brake booster. Vacuum operated systems are commonly used on light vehicles and hydraulic servos (or air brakes) on heavier vehicles. However, there are one or two exceptions. For this reason, vacuum and hydraulic servo operation is covered in this section.

Vacuum servo

A vacuum servo is used to provide braking assistance. A vacuum (low air pressure) is used to provide the extra force. The vacuum is taken from the inlet manifold on most vehicles, but a small vacuum pump can also be used. A pump is often required on diesel-engined vehicles as most do not have a throttle butterfly, and therefore do not develop any significant manifold vacuum.

The vacuum servo is fitted inbetween the brake pedal and the master cylinder. Figure 9.9 shows that the main part of the servo is the diaphragm. The larger the diaphragm, then the greater the servo assistance provided. A vacuum is allowed to act on both sides of the diaphragm when the brake pedal is in its rest position. When pedal force is applied to the piston, a bell valve cuts the vacuum connection to the rear chamber and allows air at atmospheric pressure to enter. This causes a force to act on the diaphragm so assisting with the application of the brakes.

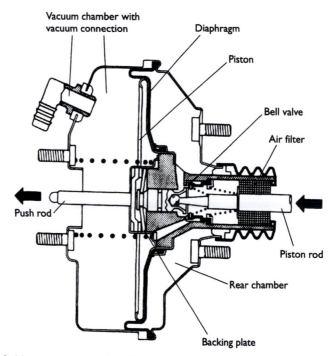

Figure 9.9 Vacuum operated brake servo

Once the master cylinder piston moves, the bell valve closes again to hold the applied pressure. Further effort by the driver on the brake pedal will open the valve again and apply further vacuum assistance. In this way, the driver can 'feel' the amount of braking effort being applied. The cycle continues until the driver effort reaches a point where the servo assistance remains fully on.

If the vacuum servo stops working, the brakes will still operate, but extra force will be required from the driver. The connection to the inlet manifold will normally be via a check valve as an extra safety feature.

Hydraulic power brakes

Hydraulic power brakes use the pressure from an engine-driven pump. This pump will often be the same one used to supply the power assisted steering. Pressure from the pump is made to act on a plunger in line with the normal master cylinder. As the driver applies force to the pedal, a servo valve opens in proportion to the force applied by the driver. The hydraulic assisting force is therefore also proportional. This maintains the important 'driver feel'.

A hydraulic accumulator (a reservoir for fluid under pressure) is incorporated into many systems. This is because the pressure supplied by the pump varies with engine speed. The pressure in the accumulator is kept between set pressures in the region of 70 bar. A warning, therefore:

If you have to disconnect any components from the braking system on a vehicle fitted with an accumulator, you must follow the manufacturer's recommendations on releasing the pressure first.

LEARNING TASKS

➡ Look back at the key words. Explain each one to a friend, and/or write out a short description to keep as evidence.

➡ Make a simple sketch to show the operation of a vacuum servo.

➡ Examine a real system and note the layout of the components.

10 Electrical systems and circuits

1 Introduction

'System' is a word used to describe a collection of related components that interact as a whole. A motorway system, the education system or computer systems are three varied examples. A large system is often made up of many smaller systems which in turn can each be made up of smaller systems and so on. Figure 10.1 shows how this can be represented in a visual form. Using the systems approach helps to split complex things into more manageable bits.

The modern motor vehicle is a very complex system. It is the ability for the motor vehicle to be split into systems on many levels which aids its design, construction and the understanding of how it works. This book has been split into vehicle systems, in the way some of the chapters are titled. In this chapter, the electrical system will be split further in to sub systems such as starting, lighting and charging.

Terminology	
System	A collection of components that carry out a function
Wiring	The method of connecting an electrical supply from one part of the vehicle to another
Battery	Source of energy in the vehicle. One of its primary jobs is to supply energy to the starter motor. The battery forms part of all the electrical systems
Charging	The alternator, when driven by the engine, is the main part of the charging system
Starting	The starter motor uses power from the battery to crank the engine when the ignition switch is operated
Lighting	All the components of a vehicle that produce light to see or be seen by! Indicators and brake lights are sometimes called auxiliaries
Instruments	'Driver information' is the modern term. Fuel and temperature gauges are the most common example
Auxiliaries	All the other electrical bits but usually things like the wipers, horns and heater blowers are thought of as auxiliaries

Open and closed loop systems

Another way of looking at a system is by considering its inputs and outputs. This method is often used to describe systems controlled by electronics such as in chapter 5 as well as this one.

An open loop system is designed to give the required output whenever a given input is applied. A good example of an open loop vehicle system would be the headlights. When the input is supplied by the switch, the output is that the headlights will come on. The feature which determines that a system is open loop is that no feedback is required for the system to operate. Figure 5.2 shows these systems in block diagram form.

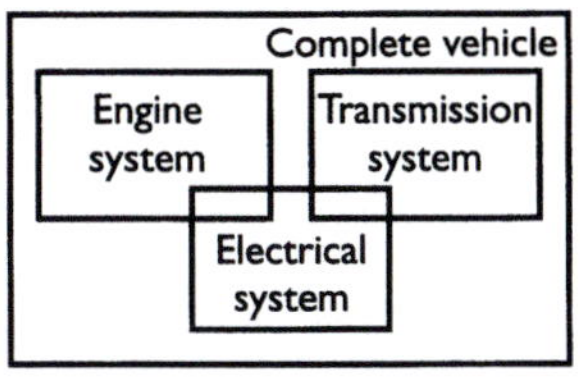

Figure 10.1 Splitting complicated things into smaller parts makes understanding easier

A closed loop system is identified by a feedback loop. It can be described as a system where there is a possibility of applying corrections if the output is not quite what is wanted. A good example of this in a vehicle is an automatic temperature control system. The interior temperature of the vehicle is determined by the output from the heater which is switched on or off in response to a signal from a temperature sensor inside the cabin. It is a feedback loop because the output from the system (temperature) is also an input to the system.

LEARNING TASKS

➡ Look back at the key words. Explain each one to a friend, and/or write out a short description to keep as evidence.

➡ Make a simple sketch to represent the lighting circuit as a block diagram system.

2 Electrical wiring

Cables and the wiring harness

KEY WORDS

■ Harness

■ CAN

Cables that are used for motor vehicle applications are now almost all copper strands insulated with PVC. Copper, beside its very low resistance, has ideal properties such as ductility and malleability (that is, it can be stretched and bent). This makes it the natural choice for most electrical conductors. PVC is again ideal as the insulation as it not only has very high resistance, but is also very resistant to petrol, oil, water and other contaminants.

The choice of cable size depends on the current drawn by the consumer (think of a small bulb compared to a large motor). The larger the cable used, the smaller the voltage loss (volt drop) in the circuit. However, the cable will be heavier. This means a trade-off must be sought between allowable volt drop and maximum cable size.

The vehicle wiring harness has developed over the years from a loom containing just a few wires, to the looms used at present on top range vehicles, containing well over a thousand separate wires. The most popular way of holding the wires together is for the bundle of cables to be spirally wrapped PVC tape. The tape is non-adhesive so as to allow the bundle of wires to retain some flexibility.

The layout of a wiring loom within the vehicle is decided after many issues are considered:

■ cable runs must be as short as possible
■ the loom must be protected against physical damage
■ the number of connections to be kept to a minimum
■ accident damage areas to be considered
■ production line techniques
■ access must be possible to main components and sub-assemblies for repair purposes.

The more connections involved in a wiring loom means that there are more areas for potential faults to develop. However, having a large multiplug assembly, which connects the engine wiring to the rest of the loom, can have advantages. During production, the engine and all its ancillaries can be fitted as a complete unit if supplied ready wired and, in the repair market, engine replacement and repairs are easier to carry out.

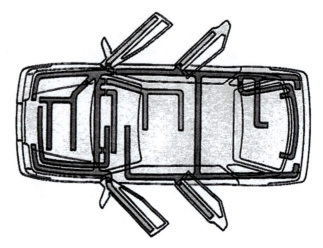

Figure 10.2 Wiring harness layout

Keeping cable runs as short as possible reduces volt drop problems and allows thinner wire to be used. This reduces the weight of the harness, which can be considerable.

The overall layout of a loom on a vehicle will broadly follow one of two patterns, that is, an 'E' shape or an 'H' shape (Figure 10.2 is like a combination of the two). The 'H' is the more common layout. It is becoming usual to have one or two main junction points as part of the vehicle wiring, these points often being part of the fuse box and relay plate.

Multiplexed wiring

The complexity of modern wiring systems has been increasing steadily over the last twenty-five years or so, and in recent years has increased dramatically. It has now reached a point where the size and weight of the wiring harness is a major problem. The number of separate wires required on a top-of-the-range vehicle can be in the region of twelve hundred. The wiring loom required to control all functions in or from the driver's door can require up to fifty wires, the systems in the dashboard area alone can use over one hundred wires and connections. This is clearly becoming a problem as, apart from the obvious issue of size and weight, the number of connections and number of wires increase the possibility of faults developing. It has been estimated that the complexity of the vehicle wiring system doubles every ten years!

The number of systems controlled by electronics is continually increasing (Figure 10.3). A number of these are already in common use and the others are becoming more widely adopted. Just some examples of these systems are listed below:

- engine management
- anti-lock brakes
- traction control
- variable valve timing
- transmission control
- cruise control
- active suspension.

All of the above systems work in their own right but are also linked to each other. Many of the sensors that provide inputs to one electronic control unit are common to some or all of the others. One solution to this is to use one computer to control all systems. A second solution is to use a common 'data bus'. This allows communication between ECU's and makes the information from the various vehicle sensors available to all of them.

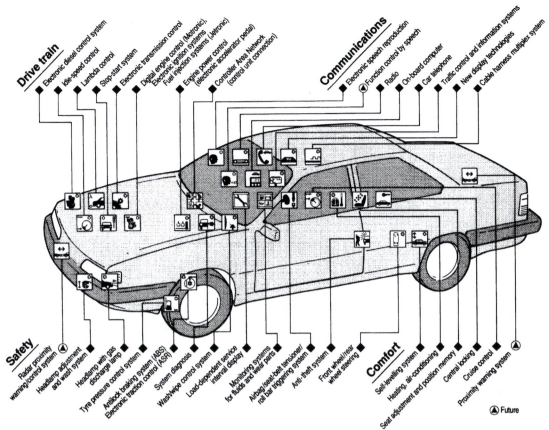

Figure 10.3 Electronics in the car

Taking this idea further, if many signals could be transmitted along one wire and made available to all parts of the vehicle, then the vehicle wiring could be reduced to just three wires. These wires would be:

- main supply
- earth connection
- signal wire.

The idea of using just one line for many signals is not new and has been in use in areas such as telecommunications for many years. Various signals can be 'multiplexed' onto one wire (fitted in, in their own time slot for example). This leads us on to the multiplexed wiring system for vehicle use, which has also been known as a ring main wiring system.

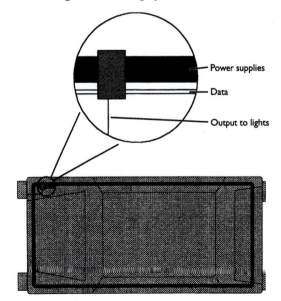

Figure 10.4 Ring main wiring

Figure 10.4 shows that the data bus and the power supply cables must 'visit' all areas of the vehicle electrical system. To illustrate the operation of this system, consider the events involved in switching the sidelights on and off. First, in response to the driver pressing the light switch, a unique signal is placed on the data bus. This signal is only recognised by special receivers built as part of each light unit assembly; these in turn will make a connection between the power ring main and the lights. The events are similar to turn off the lights except that the code placed on the data bus will be different and will be recognised only by the appropriate receivers as an 'off' code.

A system knows as Controlled Area Networks (CAN) is being introduced on some vehicles.

LEARNING TASKS

➡ Look back at the key words. Explain each one to a friend, and/or write out a short description to keep as evidence.

➡ Examine a real system and make a simple sketch to show the wiring system layout.

➡ Write a short explanation about the advantages of multiplex systems.

3 Batteries

Introduction

As a general point, the high electrical output from a battery that contains very strong chemicals makes it necessary to take precautions.

Take extra care when working with batteries

In this section we will examine the workings of the lead-acid battery, but I have also included a very short introduction to three other types.

Chemical action of the lead-acid battery

A fully charged lead-acid battery consists of the following:

- lead peroxide (PbO_2) – as the positive plates
- spongy lead (Pb) – as the negative plates
- diluted sulphuric acid ($H_2SO_4 + H_2O$) – as the electrolyte.

The dilution of the electrolyte is at a relative density of 1.28. The lead peroxide and lead are known as the active materials. A different number of electrons exist in the pure lead than when present as a compound with oxygen (lead peroxide). The voltage of a battery is created due to this difference in electrical potential (voltage). A lead-acid cell has a potential difference of about 2.1 V.

When a lead-acid cell is being charged or discharged, chemical changes take place. It can be represented by the following balanced chemical equation:

- $PbO_2 + H_2SO_4 + Pb \leftrightarrow 2PbSO_4 + 2H_2O$

In simple terms, both plates change into lead sulphate ($PBSO_4$) and the electrolyte changes to water as the battery becomes discharged.

The other reaction of interest in a battery is that of gassing after it has reached full charge. This gassing causes hydrogen and oxygen to be given off resulting in loss of water (H_2O). This is why on batteries that require topping up, you always add water and not acid. You are simply putting back what was lost.

It is the material of the grids inside a battery which cause the gassing. With sealed batteries, this has been overcome by using lead-calcium for the grid material in place of the more traditional lead-antimony.

It is accepted that the terminal voltage of a lead-acid *cell* (six cells in a 12 V battery) must not be allowed to fall below 1.75 V (10.5 V for a 12 V battery). Apart from the electrolyte tending to become very close to pure water, lead sulphate crystals grow on the plates, making it very difficult to recharge the battery.

The lead-acid battery

The battery (Figure 10.5) has a number of requirements:

■ to provide power storage and be able to release the power *quickly* to operate the vehicle starter motor

■ to allow the use of parking lights for a reasonable time

■ to allow operation of accessories when the engine is not running

■ to allow memory and alarm systems to remain active when the vehicle is left for a period of time.

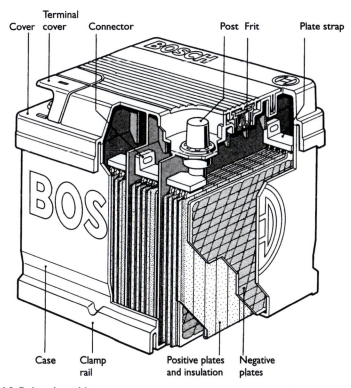

Figure 10.5 Lead-acid battery

The first two of the above list are the most important and form a major part of what is used to determine the most suitable battery for a vehicle. A further requirement of the vehicle battery is that it must be able to carry out all the functions listed above over a wide temperature range. This can be in the region of $-30°$ to $+70°C$. This is intended to cover very cold starting conditions as well as potentially high under-bonnet temperatures.

More than one hundred years of development and much research into other types of battery still leaves the lead-acid battery as the best choice. Small changes over the years have made the sealed and maintenance free battery very reliable and long lasting. This may not always appear to be the case to some end users, but note that quality is often related to the price the customer pays. Many bottom-of-the-range cheap batteries with a twelve month guarantee will last for thirteen months!

The basic construction of a 12 V lead acid battery consists of six cells connected in series. Each cell produces about 2 V and is housed in an

KEY WORDS

■ CCA

■ Active material

■ Charge rate

■ NiCad

■ Fuel cell

individual compartment within a plastic case. Figure 10.5 is a cut-away battery, showing the main parts. The active material is held in grids to form the positive and negative plates. Separators made from microporous (lots of little holes) plastic insulate these plates from each other. The grids, connecting strips and the battery posts are made from a lead alloy known as lead calcium.

Modern batteries described as 'sealed' do still have a small vent to stop the pressure build up due to gassing. An important requirement of sealed batteries is accurate control of charging voltage.

Battery ratings

In simple terms, the rating of a battery is determined by how much current it can produce and how long it sustains this current. The rate at which a battery can produce current is determined by the speed of the chemical reaction. This is determined by a number of factors:

■ surface area of the plates and quality of construction

■ quantity of active material (mostly the size of the battery)

■ temperature.

Ampere hour capacity

This is not often used, but describes how much current the battery is able to supply for 20 hours. For example, a fully charged battery quoted as being 44 Ah (Ampere-hour), will supply 2.2 A for 20 hours before being completely discharged (1.75 V per cell).

Reserve capacity

A system now used on all new batteries is reserve capacity. This is given as a time in minutes for which the battery will supply 25 A at 25°C. This is used to give an indication of how long the battery could run the car if the charging system was not working. Typically, a 44 Ah battery will have a reserve capacity of about 60 minutes.

Cold cranking amps

Batteries are given a rating to indicate performance at high current output and low temperature. A typical value of 170 A means that the battery will supply this current for one minute at a temperature of −18°C, at which point the cell voltage will fall to 1.4 V. Cold cranking capacity rating methods vary:

■ British standard is for 60 seconds

■ DIN standard and SAE standard are for 30 seconds.

Note that the *overall* power output of a battery is much greater when spread over a longer time. This is because the chemical reaction will only work at a set speed.

Maintenance and charging

By far, the majority of batteries now available are classed as 'maintenance free'. This implies that little attention is required during the life of the battery. Earlier batteries and some heavier types do, however, still require the electrolyte level to be checked and topped up periodically with distilled water. Battery posts are still a little prone to corrosion. The usual service of cleaning with hot water and the application of petroleum jelly or terminal grease is recommended. Ensuring that the battery case and, in particular, the top remains clean will help to reduce self-discharge.

The state of charge of a battery is still very important, and in general, it is not advisable to allow the state of charge to fall below 70% for long periods, as the sulphate on the plates can harden, making recharging difficult. If a battery is to be stored for a long period (more than a few weeks), then it must be recharged every so often to prevent it from becoming sulphated. Recommendations vary but a recharge every six weeks is a reasonable suggestion.

The old recommendation was that the battery should be charged at a tenth of its ampere-hour capacity for about ten hours. This figure is still valid but as ampere-hour capacity is not always used, a different method of deciding the rate is necessary. One way is to set a rate at a sixteenth of the reserve capacity, again for up to ten hours. The final suggestion is to set a charge rate at one fortieth of the cold start performance figure, also for up to ten hours. Clearly if a battery is already half charged, half the time is required to recharge to full capacity.

Boost charging is a popular technique often applied in many workshops! It is not recommended as the best method but, if correctly done and not repeated too often, is suitable for most batteries. The key to fast or boost charging is that the battery temperature should not exceed 43°C. A rate of about five times the 'normal' charge setting will bring the battery to 70–80% of its full capacity within about one hour.

The following table summarises the charging techniques for a lead-acid battery:

Constant voltage	14.4 V maximum	Will recharge any battery in seven hours or less without any risk of overcharging
Constant current	1/10 of Ah capacity	Charge time up to ten hours or less if already partly charged original state
	1/40 of cold start current	
	1/16 of reserve capacity	
Boost charging	5 times the above is the maximum	Charge can return to 70–80% of its full capacity within about an hour

Other types of batteries I have included some details about three other types of battery because you may come across them at some point. The development of electric vehicles makes it even more likely. The three types I have included are as follows:

■ alkaline batteries
■ fuel cells
■ sodium sulphur batteries.

Alkaline batteries

When a lead-acid battery is required to withstand a high rate of charge and discharge on a regular basis, or is left in a state of disuse for long periods, it will become unusable very quickly. Alkaline cells, on the other hand, require minimum maintenance and are far better able to withstand electrical abuse such as heavy discharge and over charging.

The disadvantages of alkaline batteries are that they are more bulky, have a lower energy density and are more expensive than the lead-acid equivalent. However, when the lifetime of the battery and servicing requirements are considered, the extra initial cost is worth it for some applications. Bus and coach companies and some large goods vehicle operators have used alkaline batteries. They are also well suited to electric vehicle (EV) use.

Alkaline batteries used for vehicle applications are the nickel-cadmium types. The main components of the nickel-cadmium, or nicad, cell for vehicle use are as follows:

- positive plate – nickel hydroxide NiOOH
- negative plate – cadmium Cd
- electrolyte – potassium hydroxide KOH.

The process of charging involves oxygen moving from the negative plate to the positive plate and the process is reversed when discharging. When fully charged, the negative plate becomes pure cadmium (spongy), and the positive plate becomes nickel hydrate. Cell voltage of a fully charged cell is 1.4 V but this falls rapidly to 1.3 V as soon as discharge starts. The cell is considered to be discharged at a cell voltage of 1.1 V.

Fuel cell

The energy of oxidation (burning) of conventional fuels, which is usually given off as heat, is converted directly into electricity in a fuel cell. All oxidations involve a transfer of electrons between the fuel and oxidant; this is used in a fuel cell to directly convert the energy into electricity. Battery cells involve an oxide reduction at the positive pole and an oxidation at the negative pole during some part of their chemical process. To achieve these reactions in a fuel cell, positive plates, negative plates and an electrolyte are required. The electrolyte is *fed* with the fuel.

A fuel of hydrogen when combined with oxygen proves to be an efficient design. Fuel cells are very reliable and silent in operation, but at present are very expensive to construct. Figure 10.6 shows a simplified representation of a fuel cell.

Operation of one type of fuel cell is as follows: hydrogen is passed over an electrode of porous nickel, which is coated with a catalyst, and the hydrogen diffuses into the electrolyte. This causes electrons to be stripped off the hydrogen atoms. These electrons then pass through the external circuit. Negative charge is formed at the electrode over which oxygen is passed as it also diffuses into the solution. The electrolyte is a solution of potassium hydroxide (KOH). Water is formed as the by-product of the reaction. The working temperature of these cells varies but about 200°C is typical. High pressure in the order of 30 bar is also used.

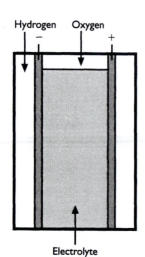

Figure 10.6 Representation of a fuel cell

Sodium sulphur

Sodium sulphur batteries have recently reached the production stage and for this reason this battery has been chosen for further discussion. The sodium sulphur or NaS battery consists of a cathode (negative plate) of liquid sodium into which is placed a current collector. This is a solid electrode of ß-alumina. A metal can surrounds the whole assembly, which is in turn in contact with the sulphur electrode anode (positive plate).

The problem with this system is that the running temperature needs to be 300–350°C. A heater rated at a few hundred watts must, therefore, form part of the charging circuit. Each cell of this battery is very small, using only about 15 g of sodium. This is a safety feature because, if the cell is damaged, the sulphur on the outside will cause the potentially dangerous sodium to be converted into polysulphides which are *comparatively* harmless. Small cells also have the advantage that they can be distributed around a car. The capacity of each cell is about 10 Ah and the output voltage of each cell is about 2 V. Figure 10.7 shows a sodium sulphur battery cell.

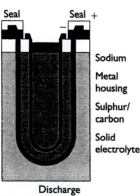

$$2Na + 3S \rightleftharpoons Na_2S_3$$

Figure 10.7 Sodium sulphur battery

This type of battery and an electric motor looks like a possible competitor to the internal combustion engine. It is estimated that the cost of running an electric vehicle will be as little as 15% of the petrol version. However, the cost of EV production is still far greater at present.

LEARNING TASKS

➡ Look back at the key words. Explain each one to a friend, and/or write out a short description to keep as evidence.

➡ Make a simple sketch to show how cells are connected in a 12 V lead-acid battery.

➡ Examine several makes of battery and note the way each is rated.

➡ Make a list of the essential steps in battery servicing.

4 Charging systems

Requirements of the charging system

KEY WORDS

- AC/DC
- Rectifier
- Regulator
- Field strength
- Warning light

The current requirement of modern vehicles for electricity is considerable, but the charging system must be able to reliably meet the demand. In simple terms, the vehicle charging system must, under all operating conditions, be able to supply all the consumers on the vehicle and still fast charge the battery.

The main part of the charging system is the alternator. Figure 10.8 shows a cut-away picture of the Bosch compact alternator. An alternator must produce DC at its output terminal as only DC can be used to charge the battery and run electronic circuits. The output of the alternator must be a constant voltage regardless of engine speed and current load.

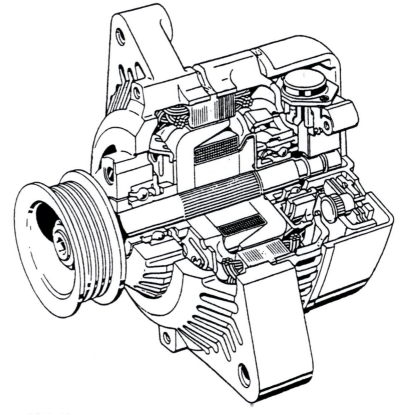

Figure 10.8 Alternator

To summarise the charging system requirements, the following criteria must be met when the engine is running:

- supply the current demands made by some or all loads
- supply whatever charge current the battery demands
- operate at idle speed
- maintain a constant voltage under all conditions
- give some indication of correct operation (e.g. warning light).

Basic principles

Figure 10.9 shows a representation of the vehicle charging system as three blocks, namely the alternator, battery and vehicle loads. When the alternator voltage is less than the battery (engine not running), the direction of current flow is from the battery to the vehicle loads. The alternator diodes prevent current flowing into the alternator. When the alternator output is greater than the battery voltage, current will flow from the alternator to both the vehicle loads and the battery. From this example, it is clear that the alternator output voltage must be above battery voltage at all times when the engine is running.

The main consideration for charging voltage is the battery voltage when fully charged. If the charging system voltage is set to this value then there can be no risk of over-charging the battery. The figure of 14 V is the accepted normal charging voltage for a 12 V system. Commercial vehicles generally employ two batteries in series at a voltage of 24 V therefore the charge voltage would be 28 V.

Leaving aside the battery and starter, the loads placed on an alternator can be considered to fall under three separate headings:

- continuous loads such as the fuel pump or ignition coil
- prolonged loads such as lights
- intermittent loads such as a heated rear window.

The charging system of a modern vehicle has to cope with these high demands under many varied conditions.

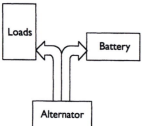

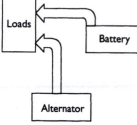

Figure 10.9 Charging system representation

Alternator operation

A rotating magnet inside stationary loops of wire causes the electromagnetic induction. In a practical alternator, the rotating magnet is an electromagnet which is supplied via two slip rings. Figure 10.10 shows the most common design, which is known as a claw pole rotor. Each end of the rotor will become a north or a south pole and hence each claw will be alternately north and south.

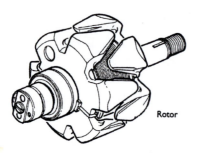

Rotor

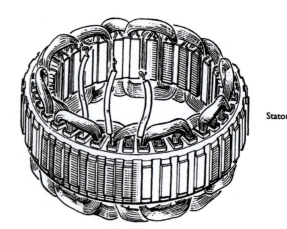

Stator

Figure 10.10 Rotor and stator

Star

Delta

Figure 10.11 Star and delta stator connections

The stationary loops of wire are known as the 'stator' and consist of three separate phases, each with a number of windings. Figure 10.10 shows a typical example. The three phase windings of the stator can be connected in two ways known as 'star' or 'delta' as shown in Figure 10.11.

In order for the output of the alternator to be used to charge the battery and run other vehicle components, it must be converted from alternating current to direct current. The component most suitable for this task is the diode. If single phase AC is passed through a diode, its output is half wave rectified as shown in Figure 10.12. In this example, the diode will only allow the positive half cycles to be conducted towards the positive of the battery. The negative cycles are blocked. A diode is often considered to be a one-way valve for electricity.

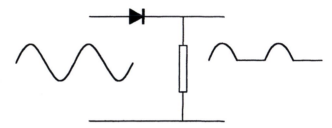

Figure 10.12 Half wave rectification

In order to rectify the output of a three-phase machine, six diodes are needed. These are connected in the form of a bridge as shown in Figure 10.13. The main part consists of three positive diodes and three negative diodes. Three further positive diodes are often included in a rectifier pack. These are usually smaller than the main diodes and are used to supply a small current back to the field windings in the rotor. They are often known as the field or excitation diodes. Due to the considerable currents flowing through the main diodes, some form of heat sink is required to prevent damage.

To prevent the vehicle battery from being overcharged, the regulated system voltage should be kept below the gassing voltage of the lead acid battery. A figure of 14.2 +/−0.2 V is typical of many, if not all 12 V charging systems. Figure 10.14 shows a voltage regulator as part of a complete alternator.

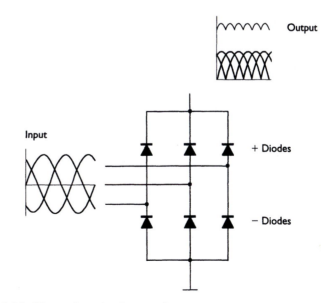

Output

Input

+ Diodes

− Diodes

Figure 10.13 Three phase bridge rectifier

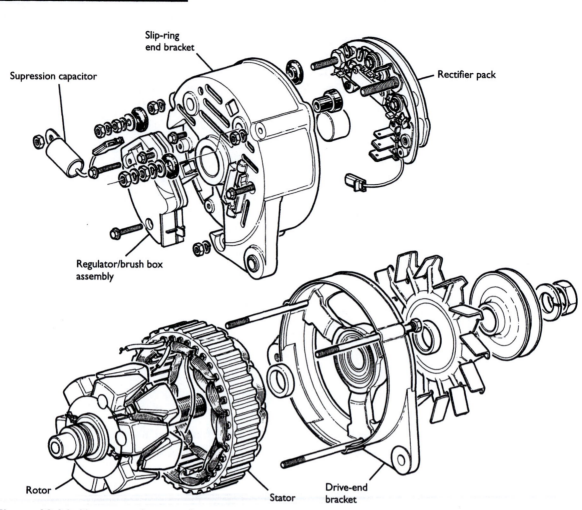

Figure 10.14 Alternator shown with its main components

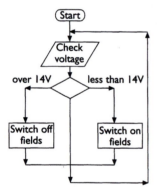

Figure 10.15 Action of a voltage regulator

Voltage regulation is a difficult task on a vehicle alternator because of the changing engine speed. The output of an alternator without regulation would rise in proportion with engine speed. Alternator output is also proportional to the magnetic field strength of the rotor and this, in turn, is proportional to the rotor current. It is the task of the regulator to control this current in response to alternator output voltage. Figure 10.15 shows a flow chart representing the action of the regulator, showing how the field current is switched off as output voltage increases and then back on again as output voltage falls. The whole switching process only takes a few milliseconds.

The key to electronic voltage regulation is the zener diode. This diode can be constructed to breakdown and conduct in the reverse direction at a precise level. This is used as the sensing element in an electronic regulator. Figure 10.16 shows a simplified electronic voltage regulator.

This regulator operates as follows: when the alternator first increases in speed, the output will be below the pre-set level. Under these circumstances, transistor T2 will be switched on by a feed to its base via resistor R3. This allows full field current to flow, thus increasing voltage output. When the pre-set voltage is reached, the zener diode will conduct. Once Z1 conducts, transistor T1 will switch on and this switches T2 off, interrupting the rotor current, causing output voltage to fall. This will cause Z1 to stop conducting. T1 will switch off, allowing T2 to switch back on and so the cycle will continue. The conventional diode D1 is to prevent the back emf from the field windings damaging the other components.

The charging circuit is one of the most simple on the vehicle in many cases. The main output is connected to the battery via suitable size cable, or in

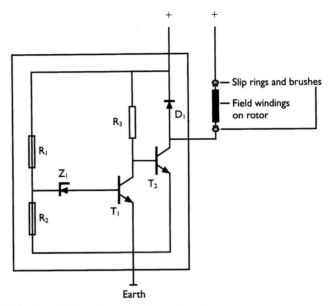

Figure 10.16 Simplified voltage regulator circuit

many cases, two cables to increase reliability and flexibility. The warning light is connected to an ignition supply on one side and to the alternator terminal at the other. A wire may also be connected to a phase terminal if it is used. Figure 10.17 shows an internal alternator circuit and its external connections.

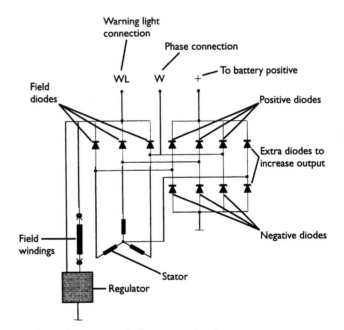

Figure 10.17 Complete internal alternator circuit

LEARNING TASKS

➡ Look back at the key words. Explain each one to a friend, and/or write out a short description to keep as evidence.

➡ Make a simple sketch to show the internal circuit of an alternator.

➡ Examine a real system and note the output of the charging system.

➡ Write a short explanation about how to change an alternator.

5   Starting systems

Introduction

KEY WORDS

- Cranking speed
- Inertia
- Pre-engaged
- Intermediate transmission
- Speed/Torque
- One-way clutch

An internal combustion engine requires the following things in order to start and continue running:

- combustible mixture
- compression of the mixture
- a form of ignition.

In order to produce the above, a minimum starting speed must be achieved. This is about 100 rev/min but a typical cranking speed is about 150 rev/min.

In comparison with many other circuits on the modern vehicle, the starter circuit is very simple. The problem to overcome, however, is that of volt drop in the main supply wires. A spring-loaded key switch, the same switch also controlling the ignition and accessories, usually operates the starter. The supply from the key switch, via a relay in many cases, causes the starter solenoid to operate and this, by a set of contacts, controls the heavy current. The basic circuit for the starting system is shown in Figure 10.18.

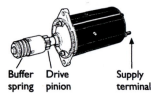

Figure 10.18 Starter circuit

The problem of volt drop in the main supply circuit is due to the high current required by the starter. This is particularly the case under adverse starting conditions such as very low temperatures.

A typical cranking current for a light vehicle engine is in the order of 150 A, but this may peak in excess of 500 A to provide the initial torque. It is generally accepted that a maximum volt drop of only 0.5 V should be allowed between the battery and starter when operating. Heavy-duty conductors (thick wires!) are therefore used.

Inertia engagement starters

In all standard motor vehicle applications, it is necessary to connect the starter to the engine ring gear only during the starting phase. If the connection remained permanent, the excessive speed at which the starter would be driven by the engine would destroy the motor almost immediately.

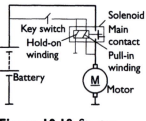

Buffer   Drive   Supply
spring   pinion   terminal

Figure 10.19 Inertia drive starter

The inertia type of starter motor has been the technique used for over eighty years, but is now becoming redundant. The starter shown in Figure 10.19 is the Lucas M35J type. It is a four pole, four brush machine and was used on small to medium petrol engined vehicles. It is capable of producing 9.6 Nm with a current draw of 350 A.

The starter engages with the flywheel ring gear by means of a small pinion. The toothed pinion and a sleeve splined onto the armature shaft are threaded such that when the starter is operated via a remote relay, the armature will cause the sleeve to rotate inside the pinion. The pinion remains still due to its inertia and, because of the screwed sleeve rotating inside it, the pinion is moved into mesh with the ring gear.

When the engine fires and runs under its own power, the pinion is driven faster than the armature shaft. This causes the pinion to be screwed back along the sleeve and out of engagement with the flywheel. The main spring acts as a buffer when the pinion first takes up the driving torque and also acts as a buffer when the engine throws the pinion back out of mesh.

One of the main problems with this type of starter is the aggressive nature of the engagement. This tends to cause the pinion and ring gear to wear. In some applications, the pinion tends to fall out of mesh when cranking due to the engine almost, but not quite, running. The pinion is also prone to seizure, often due to contamination by dust from the clutch. The pre-engaged starter motor has overcome these problems.

Pre-engaged starters

Pre-engaged starters are fitted to the majority of light vehicles in use today. They provide a positive engagement with the ring gear, as full power is not applied until the pinion is fully in mesh. They prevent premature ejection as the pinion is held into mesh by the action of a solenoid. A one way clutch is incorporated into the pinion to prevent the starter motor being driven by the engine. An example of a pre-engaged starter in common use is shown as Figure 10.20.

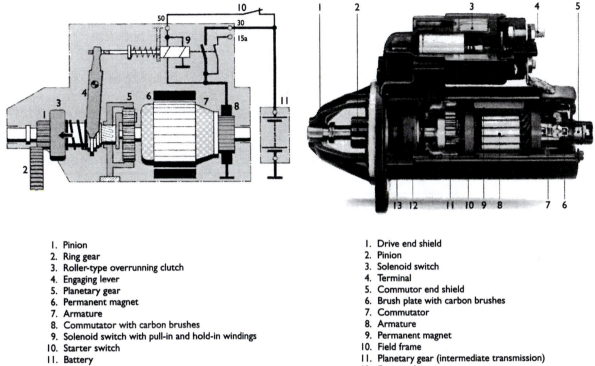

1. Pinion
2. Ring gear
3. Roller-type overrunning clutch
4. Engaging lever
5. Planetary gear
6. Permanent magnet
7. Armature
8. Commutator with carbon brushes
9. Solenoid switch with pull-in and hold-in windings
10. Starter switch
11. Battery

1. Drive end shield
2. Pinion
3. Solenoid switch
4. Terminal
5. Commutor end shield
6. Brush plate with carbon brushes
7. Commutator
8. Armature
9. Permanent magnet
10. Field frame
11. Planetary gear (intermediate transmission)
12. Engaging lever
13. Pinion-engaging drive

Figure 10.20 Pre-engaged starter and details

The basic operation of the pre-engaged starter is as follows. When the key switch is operated, a supply is made to terminal 50 on the solenoid. This causes two windings to be energised, the hold-on winding and the pull-in winding. The pull-in winding is a very low resistance and hence a high current flows. This winding is connected in series with the motor circuit and the current flowing will allow the motor to rotate slowly to facilitate engagement. At the same time, the magnetism created in the solenoid attracts the plunger and via an operating lever pushes the pinion into mesh with the ring gear. When the pinion is fully in mesh, the plunger causes a heavy-duty set of copper contacts to close. These contacts supply full battery power to the main circuit of the starter motor. When the main contacts are closed, the pull-in winding is effectively switched off due to equal voltage supply on both ends. The hold-on winding holds the plunger in position as long as the solenoid is supplied from the key switch. When the engine starts and the key is released, the main supply is removed and the plunger and pinion return to their rest positions under spring tension.

During engagement, if the teeth of the pinion hit the teeth of the flywheel (tooth to tooth abutment), the main contacts are allowed to close due to the engagement spring being compressed. This allows the motor to rotate under power and the pinion will slip into mesh.

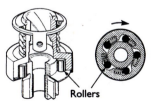

Figure 10.21 One way roller clutch drive pinion

Figure 10.21 shows a sectioned view of the one way clutch assembly. The torque developed by the starter is passed through the clutch to the ring gear. The purpose of this freewheeling device is to prevent the starter being driven at excessively high speed if the pinion is held in mesh after the engine has started. The clutch consists of a driving and driven member with several rollers in between the two. The rollers are spring loaded and either wedge-lock the two members together by being compressed against the springs, or freewheel in the opposite direction.

For applications with a higher power requirement, permanent magnet motors with intermediate transmission have been developed. This allows the armature to rotate at a higher and more efficient speed whilst still providing the torque, due to the gear reduction. The intermediate transmission, as shown in Figure 10.20, is of the epicyclic type.

LEARNING TASKS

➡ Look back at the key words. Explain each one to a friend, and/or write out a short description to keep as evidence.

➡ Make a simple sketch to show the operating mechanism of a pre-engaged starter.

➡ Examine a real system and sketch the starter circuit.

➡ Write a short explanation about how the pre-engaged starter operates.

---

6 Lighting systems

Introduction

Vehicle lighting systems are very important, particularly for road safety. If headlights were to suddenly fail at night, and at high speed, the result could be catastrophic. Most modern wiring systems fuse each bulb filament separately.

KEY WORDS

■ Alignment

■ Lens

■ GDL

■ LED

■ Dim-dip

The key point to remember with vehicle lights is that they perform two functions: to allow the driver to see, and to allow the vehicle to be seen in conditions of poor visibility. Side lights, tail lights, brake lights and others are relatively straightforward. Headlights present the most problems; namely that on dipped beam they must provide adequate light for the driver but without dazzling other road users or pedestrians. Many techniques have been tried over the years and great advances have been made, but the conflict between seeing and dazzling is very difficult to overcome.

Headlights

Light from a source such as the filament of a bulb can be projected in the form of a beam of varying patterns by using a suitable reflector and a lens. Reflectors used for headlights are usually parabolic, bifocal or homifocal. Lenses, which are also used as the headlight cover glass, are used to direct the light to the side of the road and in a downward direction. Figure 10.22 shows how lenses and reflectors can be used to direct the light. The object of the headlight reflector is to direct the random light rays produced by the bulb into a beam of concentrated light.

A reflector is basically a layer of silver, chrome or aluminium deposited on a smooth and polished surface such as brass or glass. A parabola is a curve similar in shape to the curved path of a stone thrown forward in the air. A parabolic reflector has the property of reflecting rays parallel to the principle axis when a light source is placed at its focal point, no matter where the rays fall on the reflector. It therefore produces a bright, parallel-reflected beam of constant light intensity.

As its name suggests, the bifocal reflector has two reflector sections with different focal points. This helps to take advantage of the light striking the

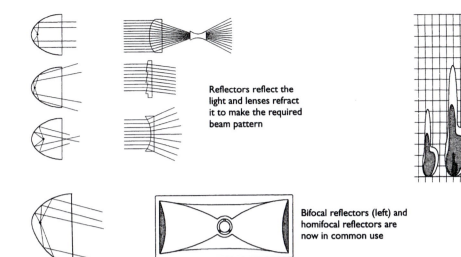

Reflectors reflect the light and lenses refract it to make the required beam pattern

Examples of improving headlight beam patterns

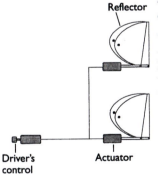

Bifocal reflectors (left) and homifocal reflectors are now in common use

Figure 10.22 Headlight patterns are produced by careful use of lenses and reflectors

lower reflector area. The parabolic section in the lower area is designed to reflect light down to improve the near field area just in front of the vehicle.

A homifocal reflector is made up of a number of sections, each with a common focal point. This design allows a shorter focal length and, hence, the light unit will have less depth. It can be used with a twin filament bulb to provide dip and main beam. The light from the main reflector section provides the normal long range lighting and the auxiliary reflectors improve near lighting.

A good headlight should have a powerful, far-reaching central beam. The light is distributed both horizontally and vertically around the beam in order to illuminate as great an area of the road surface as possible.

Headlight – levelling and aiming

The principle of headlight levelling is very simple; the position of the lights must change depending on the load in the vehicle. Figure 10.23 shows a simple levelling device operated by the driver. An automatic system can be operated from sensors positioned on the vehicles suspension. This will allow automatic compensation for whatever the load distribution on the vehicle.

Headlight adjustment is done by moving two screws positioned on the headlights. One will cause the light to move up and down, the other will cause side to side movement. Many types of beam setting equipment are available and most work on the same principle as represented by Figure 10.24. The method is the same as using an aiming board but is more convenient and accurate due to easier working and requiring less room.

Figure 10.23 Headlight levelling

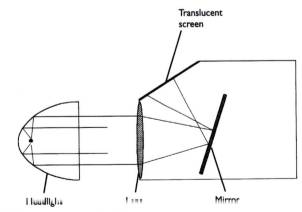

Figure 10.24 Beam setter principle

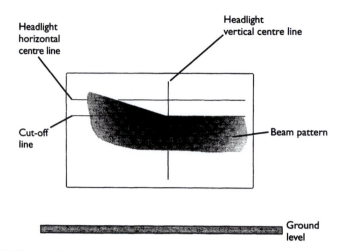

Figure 10.25 Headlight aiming board

To set the headlights of a car using an aiming board, the following procedure should be adopted:

- Park the car on level ground square on to a vertical aiming board at a distance of 10 m, if possible. Ideally, the car should be unladen except for the driver.
- Mark out the aiming board as shown in Figure 10.25.
- Bounce the suspension to ensure it is level.
- With the lights set on dip beam, adjust the cut-off line to the horizontal mark, which will be a set percentage below the height of the headlight centre, for every metre the car is away from the board. The break-off point should be adjusted to the vertical centre line for each light in turn. If the adjustment figure for the vehicle is 1.3% then the dip beam cut-off will be 1.3 cm below the horizontal line for each metre the vehicle is away from the board.

MOT testing regulations now requires the dip beam pattern to fall within set tolerances – in other words, not too high and not too low! Approved equipment must be used.

Other external lights

European and UK regulations exist relating to external lights, the following is a *simplified* interpretation of the current regulations. Typical wattage values are also given.

Sidelights	A vehicle must have two sidelights each with wattage of less than 7 W. Most vehicles have the sidelight incorporated as part of the headlight assembly
Rear or tail lights	Again, two must be fitted, each with a wattage not less than 5 W. Lights used in Europe must be 'E' marked and show a diffused light. Position must be within 400 mm from the vehicle edge, over 500 mm apart and between 350 mm and 1500 mm above the ground
Brake lights	Two lights often combined with the rear lights. They must be between 15 W and 36 W, show diffused light, and must operate when any form of first line brake is applied. Brake lights must be between 350 mm and 1500 mm above the ground and at least 400 mm apart in a symmetrical position. High level brake lights are now allowed and, if fitted, must operate with the primary brake lights

Reverse lights	No more than two lights may be fitted with a maximum wattage each of 24 W. The light must not dazzle and must either be switched automatically from the gearbox or with a switch incorporating a warning light. Safety reversing 'beepers' are now often fitted in conjunction with this circuit, particularly on larger vehicles
Day running lights	Volvo and Saab use day running lights, as these are required in Sweden and Finland. They are usually 21 W. These lights come on with the ignition and must only work in conjunction with the rear lights. Their function is to indicate that the vehicle is moving or about to move. They switch off when parking or headlights are selected
Rear fog lights	One or two may be fitted, but if only one, it must be on the offside or centre line of the vehicle. They must be between 250 mm and 1000 mm above the ground and over 100 mm from any brake light. The wattage is normally 21 W and they must only operate when either the side lights, headlights or front fog lights are in use
Front spotlights	If front spotlights (55 W is a common value) are fitted, they must between 500 mm and 1200 mm above the ground and more than 400 mm from the side of the vehicle. If the spotlights do not dip, they must only operate when the headlights are on main beam. Spot lamps are designed to produce a long beam of light to illuminate the road in the distance
Front fog lights	Front fog lamps (55 W is a common value) are fitted below 500 mm from the ground and may only be used in fog or falling snow. Fog lights are designed to produce a sharp cut-off line such as to illuminate the road just in front of the vehicle, but without reflecting back or causing glare

Figure 10.26 shows the approval markings used on headlamps and other vehicle lights.

Bulbs In the conventional bulb, the electric current heats a tungsten filament. In a vacuum, the temperature is about 2300°C. Tungsten or its alloys are ideal for use as filaments for electric light bulbs. The filament is normally wound into a 'spiralled spiral' to allow the length of thin wire into a small space and to provide some mechanical strength. If the temperature mentioned is exceeded even in a vacuum, then the filament will burn and break. This is why the voltage at which a bulb is operated must be kept within tight limits. Gas filled bulbs are more the norm where the glass bulb is filled with an inert gas such as argon. This allows the filament to work at a higher temperature without failing, and therefore produces a whiter light.

Over a period of time, some of the filament metal of normal gas bulbs evaporates and is deposited on the bulb wall. Almost all vehicles use tungsten halogen bulbs for the headlights. The bulb has a much longer life and will not blacken over a period of time as with other bulbs. The name is derived from the Greek *hal-* and *-gen*, meaning 'salt-producing'.

The glass envelope used for the tungsten halogen bulb is made from fused silicon or quartz. The tungsten filament still evaporates but, on its way to the bulb wall, the tungsten atom combines with two or more halogen atoms forming a tungsten halide. This will not be deposited onto the glass because of its temperature. Convection currents cause the halide to move back towards the filament at some point and it then splits up, returning the tungsten to the filament and releasing the halogen. Because of this, the bulb

APPROVAL MARKING

The safety critical nature of external lighting components means they are strictly controlled by both UK and European legislation and naturally fall into the scope of the M.O.T. requirements. Lamps should conform as indicated below depending on application and age of vehicle.

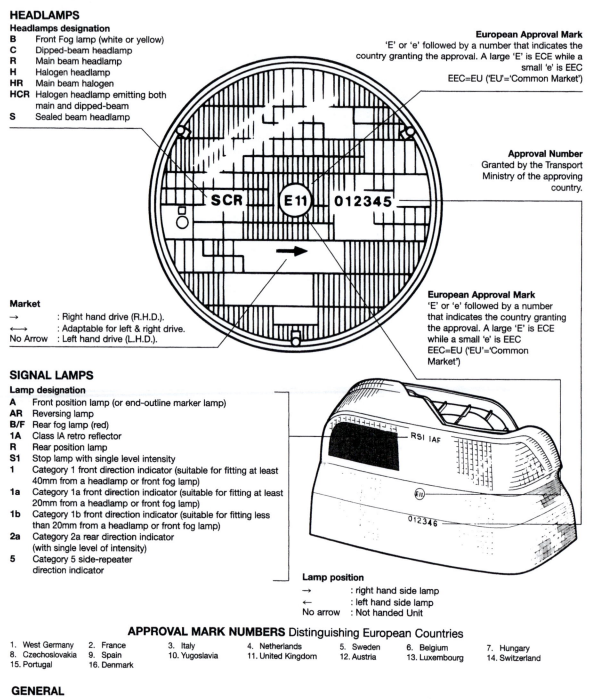

HEADLAMPS

Headlamps designation

B	Front Fog lamp (white or yellow)
C	Dipped-beam headlamp
R	Main beam headlamp
H	Halogen headlamp
HR	Main beam halogen
HCR	Halogen headlamp emitting both main and dipped-beam
S	Sealed beam headlamp

European Approval Mark
'E' or 'e' followed by a number that indicates the country granting the approval. A large 'E' is ECE while a small 'e' is EEC
EEC=EU ('EU'='Common Market')

Approval Number
Granted by the Transport Ministry of the approving country.

Market

→	: Right hand drive (R.H.D.).
←→	: Adaptable for left & right drive.
No Arrow	: Left hand drive (L.H.D.).

European Approval Mark
'E' or 'e' followed by a number that indicates the country granting the approval. A large 'E' is ECE while a small 'e' is EEC
EEC=EU ('EU'='Common Market')

SIGNAL LAMPS

Lamp designation

A	Front position lamp (or end-outline marker lamp)
AR	Reversing lamp
B/F	Rear fog lamp (red)
1A	Class IA retro reflector
R	Rear position lamp
S1	Stop lamp with single level intensity
1	Category 1 front direction indicator (suitable for fitting at least 40mm from a headlamp or front fog lamp)
1a	Category 1a front direction indicator (suitable for fitting at least 20mm from a headlamp or front fog lamp)
1b	Category 1b front direction indicator (suitable for fitting less than 20mm from a headlamp or front fog lamp)
2a	Category 2a rear direction indicator (with single level of intensity)
5	Category 5 side-repeater direction indicator

Lamp position

→	: right hand side lamp
←	: left hand side lamp
No arrow	: Not handed Unit

APPROVAL MARK NUMBERS Distinguishing European Countries

1. West Germany	2. France	3. Italy	4. Netherlands	5. Sweden	6. Belgium	7. Hungary
8. Czechoslovakia	9. Spain	10. Yugoslavia	11. United Kingdom	12. Austria	13. Luxembourg	14. Switzerland
15. Portugal	16. Denmark					

GENERAL

RH and LH vehicle lamp positions are as viewed from the drivers seat. The photographic representation of the Unipart replacement headlamp/indicator, is intended to show the specification of the part supplied. Generally only the RH (offside) unit is pictured. Units are supplied less bulbs.

Figure 10.26 Approval marking

will not become blackened and the light output will remain constant throughout its life. The envelope and the filament can also be made smaller, allowing better focusing.

Lighting circuits   Figure 10.27 shows a simplified lighting circuit. Whilst this representation helps to show the way in which a lighting circuit operates, it is not now used in this simple form. The circuit does help, however, to show in a simple way how various lights in and around the vehicle operate with respect to each other. For example, fog lights can be wired so they will only work when the sidelights are on. Another example is how the headlights can not be switched on without first switching on the sidelights.

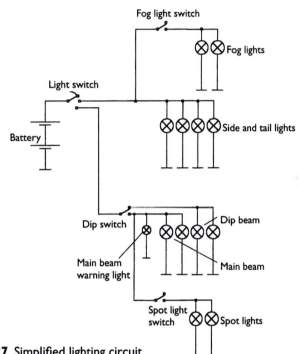

Figure 10.27 Simplified lighting circuit

Dim dip headlights are an attempt to stop drivers just using sidelights in semi-dark or poor visibility conditions. The circuit works so that when sidelights and ignition are on together, the headlights will come on automatically at about one sixth of normal power. Dim dip lights are usually achieved by using a simple resistor in series with the headlight bulb.

Figure 10.28 shows the complete lighting system of a light vehicle. Operation of the main parts of this circuit is detailed below.

Sidelights

Operation of the switch (marked as 634) allows the supply on the N or N/S wire (British standard colour codes are discussed on page 188 of the level two book), to pass to fuses 7 and 8 on an R wire. The two fuses then supply left sidelights and right sidelights as well as the number plate light.

Dipped beam

When the dip beam is selected, a supply is passed to fuse 9 on a U wire and then to the dim dip unit which is now de-energised. This allows a supply to fuses 10 and 11 on the O/U wire. This supply is then passed to the left light on a U/K wire and the right light on a U/B wire.

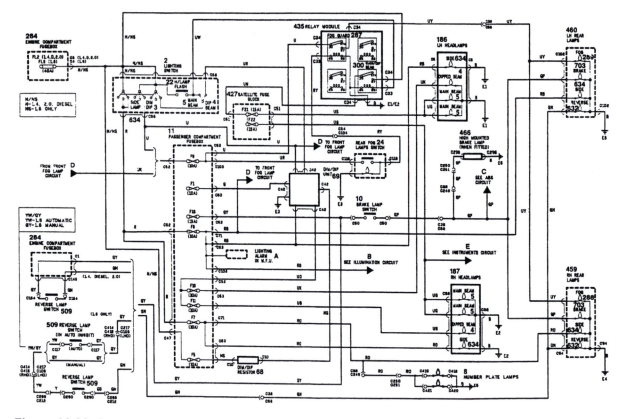

Figure 10.28 Complete vehicle exterior lighting circuit

Main beam

Selecting main beam allows a supply on the U/W wire to the main/dip relay, thus energising it. A supply is therefore placed on fuses 21 and 22 and hence to each of the headlight main beam bulbs.

Dim dip

When sidelights are on, there is a supply to the dim dip unit on the R/B wire. If the ignition supplies a second feed on the G wire, then the unit will allow a supply from fuse 5 to the dim dip resistor on an N/S wire and then on to the dim dip unit on an N/G wire. The unit then links this supply to fuses 10 and 11 (dip beam fuses).

Figure 10.29 shows two common auxiliary lights, a fog and spot light.

Figure 10.29 Auxiliary lights

Gas discharge lamps

Gas discharge lamps (GDL) have the potential to provide more effective illumination and new design possibilities for the front of a vehicle. The GDL system consists of three main components, listed in the following table.

Lamp	This operates in a very different way from conventional incandescent bulbs. A much higher voltage is needed. Figure 10.30 illustrates the operating principle of a GDL
	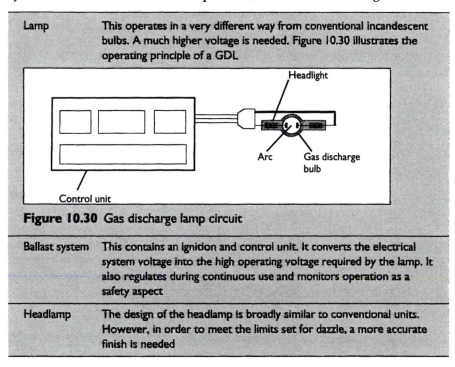
	Figure 10.30 Gas discharge lamp circuit
Ballast system	This contains an ignition and control unit. It converts the electrical system voltage into the high operating voltage required by the lamp. It also regulates during continuous use and monitors operation as a safety aspect
Headlamp	The design of the headlamp is broadly similar to conventional units. However, in order to meet the limits set for dazzle, a more accurate finish is needed

The source of light in the gas discharge lamp is an electric arc. The actual discharge bulb (made from quartz glass) is only about 10 mm across and contains two electrodes, 4 mm apart. The distance between the end of the electrode and the bulb contact surface is 25 mm; this corresponds with the dimensions of the standard H1 bulb.

At room temperature, the bulb contains a mixture of mercury, various metal salts and xenon under pressure. When the light is switched on, the xenon illuminates at once and evaporates the mercury and metal salts. The high luminous efficiency is due to the metal vapour mixture. The mercury generates most of the light and the metal salts affect the colour. The high output of UV radiation from the GDL means that, for safety reasons, special filters are required. The average output of the GDL is three times greater than a standard bulb.

Ultra-violet headlights

Since UV radiation is virtually invisible, it cannot dazzle oncoming traffic but will illuminate fluorescent objects such as specially treated road markings and clothing. They glow in the dark much like a white shirt under some disco lights. The UV light will also penetrate fog and mist, as the light reflected by water droplets is invisible. It will even pass through a few centimetres of snow.

Four-headlamp cars are being used consisting of two conventional halogen dip lights and two UV lights. The UV lights come on at the same time as the dipped beams, effectively doubling their range but without dazzling.

GDL's are used as the light source and two stage blue filters are used to eliminate visible light. Precise control of the filter colour is needed to ensure UVB and UVC are filtered out as these can cause eye damage and skin cancer. This leaves UVA, which is just beyond the visible spectrum. Some danger still exists, for example, if a child was to look directly and at close range into the faint blue glow of the lights, eye damage could result. To prevent this, the lights will only operate when the vehicle is moving.

LED lighting

Until recently, legislation prevented the use of LED's for exterior lighting. The advantages of LED lighting are clear, the greatest being reliability. LED's have a typical rated life of over 50 000 hours, compared to just a few thousand for normal bulbs. The environment in which vehicle lights have to survive consists of extreme variations in temperature and humidity as well as serious shocks and vibration.

LED's are more expensive than bulbs but the potential savings in design costs, due to sealed units being used and greater freedom of design, will outweigh the extra expense. A further advantage is that they turn on quicker than ordinary bulbs. This time is approximately the difference between 130 ms for the LED's, and 200 ms for bulbs. If this is related to a vehicle brake light at motorway speeds, then the increased reaction time equates to about a car length. This is potentially a major contribution to road safety.

The reliability of the LED allows designers to integrate the lights into the vehicle body in ways which have so far not been possible. The colour of light emitted by an LED is red, orange, amber, yellow or green. Developments are under way to produce a blue LED which, when combined with red and green, will allow white light from a solid state device. This will work in much the same way as the combinations of pixels on a colour television screen.

Their shock resistance will allow LED's to be mounted on the boot lid. Many manufacturers are designing rear spoilers with brake lights built in, which is an advantage for safety, as well as looking good.

LEARNING TASKS

➡ Look back at the key words. Explain each one to a friend, and/or write out a short description to keep as evidence.

➡ Make a simple sketch to show a headlight aligner in use.

➡ Examine the lighting circuit shown above and follow each light operation in turn.

➡ Write a short explanation about the benefits of UV lights.

---

7 Auxiliary circuits

Windscreen washers and wipers

The wiper system must meet the following requirements:

- efficient removal of water and snow
- efficient removal of dirt
- high and low temperature operation
- long lasting.

KEY WORDS

- Flash rate
- Warning light
- Third brush
- Flasher unit
- Off screen parking
- Wash wipe

The wiper blades are made of a rubber compound and are held onto the screen by a spring in the wiper arm. The aerodynamic properties of the wiper blades has become increasingly important to the design of the vehicle, as different air currents flow on and around the screen area. The lip wipes the surface of the screen at an angle of about 45°.

Most wiper motors now in use are permanent magnets, have three brushes and drive via a worm gear to increase torque. The three brushes allow two-speed operation. The normal speed operates through two brushes placed in the usual positions opposite to each other. For a fast speed, the third brush is placed closer to the earth brush. This reduces the number of armature

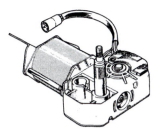

Figure 10.31 Wiper motor

windings between them, which increases current and, therefore, speed. Figure 10.31 shows a typical wiper motor. Typical specifications for wiper motor speed and, hence, wipe frequency are 45 rpm at normal speed and 65 rpm at fast speed. Figure 10.32 shows a typical wiper linkage.

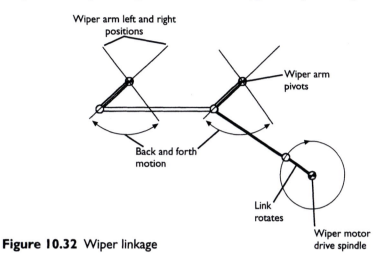

Figure 10.32 Wiper linkage

The windscreen washer system consists of a simple permanent magnet motor driving a centrifugal water pump. The water, preferably with a cleaning additive, is directed onto the screen by two or more jets. A non-return valve is often fitted in the line to the jets to prevent water siphoning back to the reservoir. This also allows 'instant' operation when the washer button is pressed. The washer circuit is normally linked in to the wiper circuit so that the wipers start automatically when the washers are operated and will continue for several sweeps after the washers have stopped.

Figure 10.33 shows the circuit of a programmed wiper system. The ECU contains two relays to enable the motor to be reversed. Further control of

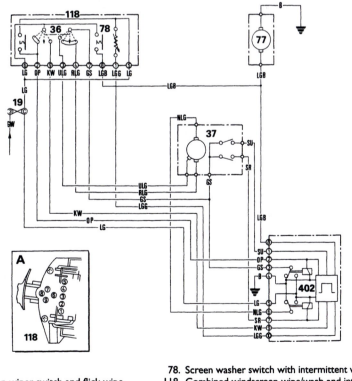

19. Fuse box
36. Windscreen wiper switch and flick wipe
37. Windscreen wipe motor
77. Screen washer pump

78. Screen washer switch with intermittent wiper control
118. Combined windscreen wipe/wash and intermittent wiper switch
402. Windscreen wipe/wash programmed control
A. Switch connectors – wipe/wash and intermittent wipe only

Figure 10.33 Column switch – wash wipe with programmed washer wipe and variable intermittent wipe circuit

wipers other than just delay is possible with electronic control. The system consists of a two-speed motor with two limit switches, one for the park position and one which operates at the top limit of the sweep. A column switch is usually used, which has positions for wash/wipe, fast speed, slow speed, flick wipe and delay, which has several further settings. The heart of this system is the programmed wiper control unit. An interesting feature is that the wiper blades are parked below the screen. This is achieved by using the top limit switch to signal the ECU to reverse the motor for parking. The switch is normally closed and switches open circuit when the blades reach the 'A' post. Due to the design of the linkage, the arms move further when working in reverse and pull the blades off the screen. The normal park limit switch stops the motor via the ECU in this position.

Indicator circuits   Direction indicators have a number of statutory requirements:

- The colour must be amber but they may be grouped with other lamps.
- The flashing rate must be between one and two per second.
- If a fault develops, this must be apparent to the driver by the operation of a pilot light on the dashboard. The fault can be indicated by a distinct change in frequency of operation or by the pilot light remaining on.
- If one of the main bulbs fails, then the remaining lights should continue to flash perceptibly.
- The wattage of indicator light bulbs is normally 21 W.

Flasher units capable of operating different numbers of bulbs, such as for when towing a trailer or caravan, are available. Most units use a relay for the actual switching as this provides an audible signal.

The circuit diagram shown in Figure 10.34 shows the full layout of the indicator and hazard lights wiring. Note how the hazard switch, when

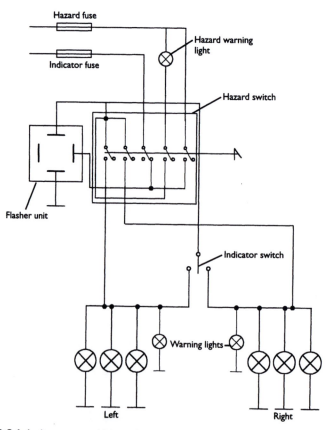

Figure 10.34 Indicators and hazard circuit

operated, disconnects the ignition supply from the flasher unit and replaces it with a constant supply. The hazard system will, therefore, operate at any time but the indicators will only work when the ignition is switched on. When the indicator switch is operated left or right, the front, rear and repeater bulbs are connected to the output terminal of the flasher unit, which then operates and causes the bulbs to flash.

When the hazard switch is operated, five sets of contacts are moved. Two sets connect left and right circuits to the output of the flasher unit. One set disconnects the ignition supply and another set connects the battery supply to the unit. The final set of contacts cause a hazard warning light on the dashboard to be operated.

LEARNING TASKS

➡ Look back at the key words. Explain each one to a friend, and/or write out a short description to keep as evidence.

➡ Examine a real system and make a simple sketch to show the wiper linkage.

➡ Write a short explanation about how to choose the correct flasher unit for a vehicle.

8 Instrumentation

Sensors   Sensors are used in vehicle applications for many purposes. For example, the coolant temperature thermistor is used to provide data to the engine management system as well as to the driver, via some type of display. For the purpose of providing information to the driver via the vehicle instruments, the following list of variables, together with typical sensors, is representative of today's vehicles.

KEY WORDS

■ Sensors

■ VCM

■ LED

■ VFD

■ HUD

■ LCD

■ Thermal gauge

Variable	Sensor example
Fuel level	Variable resistor
Temperature	Thermistor
Bulb failure	Reed relay
Road speed	Inductive pulse generator
Engine speed	Hall effect
Fluid levels	Float and reed switch
Oil pressure	Diaphragm switch
Brake pad wear	Embedded contact wire
Lights in operation	Bulb and simple circuit
Battery charge rate	Bulb circuit/voltage monitor

Figure 10.35 shows some of the sensors listed above.

Thermal type gauges   Thermal gauges, primarily used for fuel and engine temperature indication, have been used for many years and will continue to be used, due to their simple design. The gauge works by using the heating effect of electricity on a bimetal strip (Figure 10.36). As a current flows through a heating coil wound on a bimetal strip, the heat causes the strip to bend. The bimetal strip is connected to a pointer on a scale. The amount of bend is proportional to the

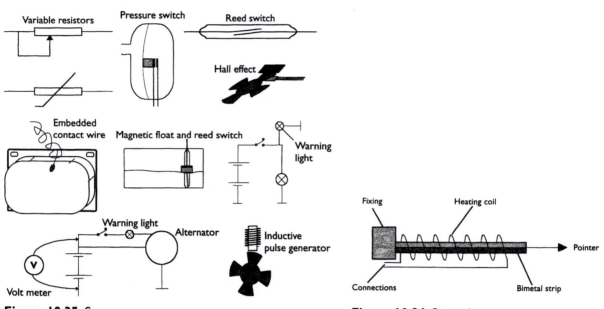

Figure 10.35 Sensors

Figure 10.36 Bimetal strip operation

heat, which, in turn, is proportional to the current flowing. Providing the *sensor* can vary its resistance in proportion to the variable being measured (fuel level for example), the gauge will indicate a value on a scale.

The needle moves very slowly to its final position because of the slow thermal effect on the bimetal strip. This is a particular advantage for fuel level display, as the variable resistor in the tank will move as the fuel moves due to vehicle movement. If the gauge were able to react quickly it would be constantly moving. The movement of the fuel is averaged out and a relatively accurate display can be obtained.

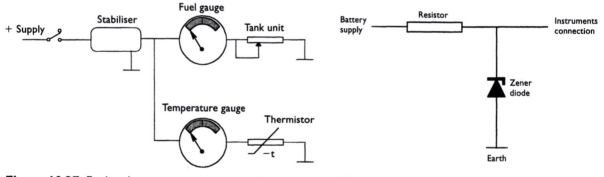

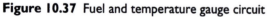

Figure 10.37 Fuel and temperature gauge circuit

Figure 10.38 A voltage stabiliser

Thermal type gauges are used with a variable resistor and float in a fuel tank or with a thermistor in the engine water jacket. Figure 10.37 shows the circuit of these two together. Note that a constant voltage supply is required to prevent changes in the vehicle system voltage affecting the reading. This is because if system voltage increased, the current flowing would increase and hence the gauges would read higher. Most voltage stabilisers are simple zener diode circuits, represented by Figure 10.38.

Other types of gauges Gauges to display road or engine speed need to react very quickly to changes. Some systems use stepper motors for this purpose, although many retain the conventional cable driven speedometers. Figure 10.39 shows a block diagram of a speedometer system, which uses an ammeter as the gauge.

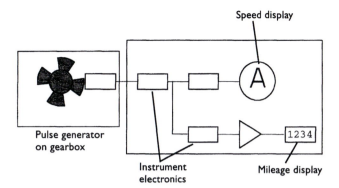

Speed display

Pulse generator
on gearbox

Instrument
electronics

Mileage display

1234

Figure 10.39 Electronic speedometer

This system uses a sensor, which will produce a constant voltage signal even
at very low speed. The frequency of the signal is proportional to road speed.
The sensor is driven from the gearbox or final drive output. The gauge will
read the average of the pulses. A system for operating a tachometer is very
similar to the speedometer system. Pulses from the ignition primary circuit
are often used to drive this gauge.

An instrumentation system

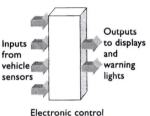

Inputs
from
vehicle
sensors

Outputs
to displays
and
warning
lights

Electronic control
unit

Figure 10.40
Instrumentation system

The block diagram shown as Figure 10.40 is a representation of a modern
instrumentation system. All signal conditioning and other functions are
carried out in the ECU. This will now often form part of the dashboard
assembly. Sensors provide information to the ECU, which in turn will drive
suitable displays or gauges. The ECU contains memory, which allows it to be
programmed to a specific vehicle. Some of the extra functions available with
this system are described briefly below.

- Low fuel warning light – can be made to illuminate at a particular
resistance reading from the fuel tank sender unit.
- High engine temperature warning light – can be made to operate at a set
resistance of the thermistor.
- Oil pressure or other warning lights can be made to flash – this is more
likely to catch the driver's attention.
- Service or inspection interval warning lights can be used – BMW are
well known for this. The warning lights are operated broadly as a
function of time but, for example, the service interval is reduced if the
engine experiences high speeds and/or high temperatures.

Vehicle condition monitoring

VCM or vehicle condition monitoring is a form of instrumentation. Figure
10.43 shows an instrument display unit, which also incorporates the vehicle
map. The complete VCM system can include driver information relating to
the following list.

Systems which can be monitored:

- high engine temperature

Figure 10.41 The instruments feature indirect green illumination, which has
proven to be better than red in night driving. The graphic pictogram and warning
symbols are normally blanked

- low fuel
- low brake fluid
- worn brake pads
- low coolant level
- low oil level
- low screen washer fluid
- low outside temperature
- bulb failure
- doors, bonnet or boot open warning.

Shown as Figure 10.42, a circuit can be used to operate bulb failure warning lights. The principle is that the reed relay is only operated when the bulb being monitored is drawing current. Oil level monitoring can be by measuring the resistance of a heated wire on the end of the dipstick. A small current is passed through the wire to heat it. How much of the wire is covered by oil will determine its temperature and, therefore, its resistance.

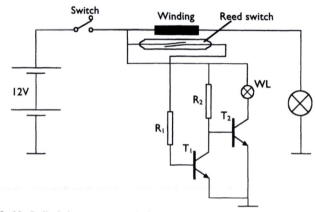

Figure 10.42 Bulb failure warning light circuit

Many of the circuits monitored use a dual resistance system so that the circuit itself is also checked. The high or low resistance readings are used to indicate, for example, correct fluid level and low fluid level. A figure outside these limits would indicate a circuit fault of either a short or open circuit connection.

The display is often just a collection of LED's or a back lit liquid crystal display (LCD). These are arranged into suitable patterns and shapes to represent the circuit or system being monitored. An open door will illuminate a symbol, which looks like the door of the vehicle map (plan view of the car) is open. Low outside temperature or ice warning is often a large snowflake!

Trip computer   The trip computer used on many top range vehicles is arguably an expensive novelty, but is popular none-the-less. Figure 10.43 shows the display of a typical trip computer. The functions available on most systems are listed below.

- time and date
- elapsed time or a stop watch
- estimated time of arrival
- average fuel consumption
- range on remaining fuel
- trip distance.

The above details can usually be displayed in imperial or metric units. In order to calculate the above outputs, the following inputs to the system are required.

- clock signal from a crystal oscillator
- vehicle speed from the speed sensor or instruments ECU

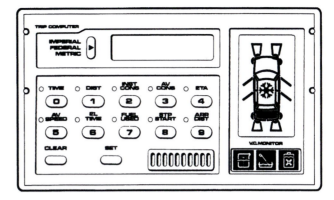

Figure 10.43 A trip computer and vehicle map

■ fuel being used from injector open time or a flow meter
■ fuel in the tank from the tank sender unit
■ Mode/Set/Clear from input by the driver.

Other systems Further systems that could be said to form part of the instrumentation are as follows:

■ distance from vehicle in front
■ distance behind the vehicle when reversing
■ traffic information
■ road maps and position indication
■ route finding.

The first two in this list rely on some form of radar range finding. A radio signal is transmitted from the vehicle. The time for the signal to return after reflecting off an obstruction gives an accurate indication of the distance.

General road and traffic information is becoming increasingly important. Vehicles have been equipped with a device which helps drivers get to a destination by displaying their vehicle's location on a map. Data from a solid-state compass installed in the vehicle's roof and from sensors mounted on its wheels are processed by a computer and displayed on a dashboard screen. The car's position is represented as a fixed triangle on a map, which scrolls down as the car moves forward and rotates sideways when it turns. Satellite positioning systems are also being used, as are systems that transmit data to the vehicle from roadside units.

Jaguar in conjunction with other manufacturers, as part of a project called 'Prometheus', has developed a computerised system that picks up information from static transmitters. This system gives directions and advance warning of road junctions, sign posts and speed limits etc.

Other forms of driver information systems are being considered, such as one which is the electronic equivalent of winding down a window and asking for directions. By choosing from screen menus, the driver can specify where s/he wants to go. Twenty seconds later, a printed sheet of driving instructions constructed from a cartographic database will be printed. Computerised route finding software is already very popular. Its one problem is that the data on disk is out of date instantly due to road works and other restrictions. Transmitting live data to the vehicle is the way to avoid this problem.

Instrument displays   *LED's*

When the junction of a diode is manufactured in a certain way, light will be emitted when a current is made to pass in the forward direction. This is a light emitting diode (LED) and will produce red, yellow, or green light with slight changes in the manufacturing process. LED's are used extensively as indicators on electronic equipment and in digital displays. They last for a very long time (50 000 hours) and draw only a very small current. LED displays are tending to be replaced, for automobile use, by the liquid crystal type, which can be back lit to make it easier to read in the daylight. However LED's are still popular for many applications.

The actual display will normally consist of a number of LED's arranged into a suitable pattern for the required output. This can range from a seven-segment display to show numbers, to a custom designed speedometer display. Figure 10.44 shows a display made up of LED's.

Figure 10.44 A display using LED's

LCD's

Liquid crystals are substances that do not melt directly to the liquid phase, but first pass through a stage in which the molecules are partially ordered. In this stage, a liquid crystal is a cloudy or translucent fluid but still has some of the optical properties of a solid crystal.

The molecular structure of liquid crystals can be altered easily by electrical voltages. A liquid crystal also scatters light that shines on it. Because of these properties, liquid crystals are often used to display letters and numbers on calculators, digital watches and automobile instrument displays. This display is achieved by only allowing polarised light to enter the liquid crystal which, as it passes through the crystal, is rotated by ninety degrees. The light then passes through a second polariser, which is set at ninety degrees to the first. A mirror at the back of the arrangement reflects the light so that it returns through the polariser, the crystal and the front polariser again. The net result is that light is simply reflected, but only when the liquid crystal is in this one particular state.

When an AC voltage is applied to the crystal, it becomes disorganised and the light passing through it is no longer twisted by ninety degrees. This

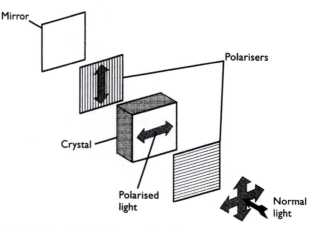

Figure 10.45 Principle of a liquid crystal display

means that the light polarised by the first polariser will not pass through the second, and will therefore not be reflected. This will show as a dark area on the display. These areas are constructed into suitable segments in much the same way as with LED's to provide whatever type of display is required. Figure 10.45 shows a representation of how a liquid crystal display works. LCD's are very low power, but they do require an external source of light to operate. To be able to read the display in the dark, some form of lighting for the display is required.

VFD's

A vacuum fluorescent display works in much the same way as a television tube and screen (Figure 10.46). It is popular for vehicle use because it can produce a bright light (which is adjustable), and a wider choice of colours than LED or LCD displays. The VFD system consists of three main components. These are the filament, the grid and the screen, with segments placed appropriately for the intended use of the display. The control grid is used to act as a brightness control as the voltage is altered.

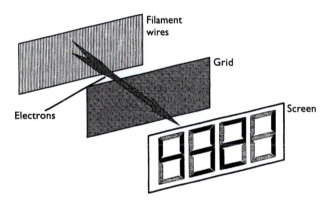

Figure 10.46 Vacuum fluorescent display

When a current is passed through the tungsten filaments, they become red-hot and emit electrons. The segments making up the display are coated with a fluorescent substance and connected to a wire. The segments are given a positive voltage, attracting electrons. When electrons strike the segments, they fluoresce, emitting a yellow-green or a blue-green light, depending on the type of coating.

The glass front of the display can be coloured to improve the readability. These displays have many advantages but the main problem for automobile use is a susceptibility to shock and vibration. This can be overcome, however, with suitable mountings.

LEARNING TASKS

➡ Look back at the key words. Explain each one to a friend, and/or write out a short description to keep as evidence.

➡ Examine a real system and make a simple sketch to show the layout of the instrumentation displays.

➡ Write a short explanation about why slow operation of the fuel gauge is a good point but slow operation of the temperature gauge is not.

9   Electric vehicles

Introduction   During the last decade or so the pressure to produce a non-fossil fuel vehicle has increased. Legislation is set to require the production of zero emission vehicles (ZEV's) in the very near future.

Even as far back as 1990, General Motors announced that its EV, the 'Impact' could accelerate to 100 km/h in just eight seconds, had a top speed of 160 km/h (100 mph), and had a range of 200 km between charging. A totally new design from its drag-reducing tyres to its brakes, which, when engaged, act as small electric generators, the car is powered by a 397 kg array of advanced gel electrolyte lead-acid batteries. Two small AC electric motors drive the front wheels. Impact's range, however, is only 240 km before the batteries must be recharged. The recharging time is about two hours; this can be reduced to one hour in an emergency, but has the effect of reducing battery life.

A number of options are available when designing an electric car but, at the risk of over simplification, the most important choices are as follows:

- What type of batteries?
- What type of drive system?

The great advantage of lead acid batteries is the existing mature technology, which is accepted by the motor industry. The disadvantage is the relatively low power to weight ratio. The sodium sulphur battery seems to be a good contender, but has a far greater cost, and new technologies are needed to cope with the operating conditions such as the high temperatures.

The second question relating to the type of drive used again offers a number of choices. The basic choice is between an AC and DC motor. The AC motor offers control advantages but requires the DC produced by the batteries to be converted using an inverter. A DC shunt wound motor rated at about 45 kW appears to be one of the popular choices for the smaller vehicles.

Drive motors   Several type of motor are either in use or are serious contenders.

AC motors – general	In general, all AC motors work on the same principle. A three-phase winding is distributed round a laminated stator and sets up a rotating magnetic field. This rotating field causes a rotor to move
Asynchronous motor	The asynchronous motor is often used with a squirrel cage rotor made up of a number of pole pairs. The rotating magnetic field in the stator induces an emf in the rotor which, because it is a complete circuit, causes current to flow. This creates magnetism, which reacts to the original field caused by the stator, and, hence, the rotor rotates
Synchronous with permanent excitation	This motor has a wound rotor known as the inductor, which is magnetised by a DC supply. The magnetism 'locks on' to the rotating magnetic field and produces a constant torque. The motor needs special arrangements for starting rotation. An advantage, however, is that it makes an ideal generator. The normal vehicle alternator is very similar
EC motors (electronically controlled)	The EC motor is, in effect, half way between an AC and a DC motor. Its principle is very similar to the synchronous motor above except the rotor contains permanent magnets and, hence, no slip rings. It is sometimes known as a brushless motor. A control system produces a rotating field, the frequency of which determines motor speed
DC motors	The DC motor is a well-proven device and has been used for many years on electric vehicles such as milk floats and forklift trucks. Its main disadvantage is that the high current has to flow through the brushes and commutator

EV summary   The concept of the electric vehicle is not new; the battery technology was developed in the late 19th century and many cars were being manufactured by the turn of the year 1900! Although some models achieved high speeds at that time, the electric car was generally slow and expensive to operate. Its range was also limited by its dependence on facilities to recharge the battery. Many of these problems have been overcome, but not all of them, notably the cost. It is important to note, however, that 'cost' is a relative value and when the consequences of pollution are considered, the cost may not be as high as it appears. Advances in batteries and other technologies keep increasing the range and performance of the EV. Watch this space!

Hybrid vehicles   The concept of a combined power source vehicle is simple. Internal combustion engines produce dangerous emissions and have poor efficiency, particularly at part load. Electric drives produce 'no' dangerous emissions, but have a limited range. The solution is to combine the best aspects of both, and minimise the worst. Such is the principle of the hybrid drive system.

One way of using this type of vehicle is to use the electric drive in slow traffic and towns, and to use the internal combustion engine on the open road. This could be the most appropriate way for reducing pollution in the towns. Sophisticated control systems actually allow even better usage, so that under certain conditions, the electric motor and the engine can be used together.

The hybrid or combined power source vehicle is likely to become popular. It appears to be an ideal and obvious compromise, whilst the EV drive and battery technology is developing. It may become possible in the future to produce a fossil fuel engine which, when running at a constant speed, will produce a level of emissions, if not zero then very close. It has now become accepted that there will be no miracle battery, at least in the foreseeable future. The energy density of fossil fuels is an order of magnitude above any type of battery. This gives credence to the hybrid design.

LEARNING TASKS

➡ Look back at the key words. Explain each one to a friend, and/or write out a short description to keep as evidence.

➡ Read about other types of EV's and batteries in another book. Will the electric vehicle take over in the future?

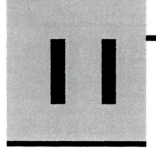

Comfort and safety

1 Introduction

Start here!

Comfort and safety is an unusual heading but it serves to catch all the bits that would not fit anywhere else! Safety, of course, is a serious concern. A great deal of time and money is spent by the vehicle manufacturers trying to ensure that the vehicles we drive are as safe as possible. Our job is to ensure that they stay that way. Comfort in a vehicle is a contribution to safety, but some systems such as electric seats and windows are simply to make the vehicle more pleasant to use. In-car entertainment (ICE) systems could have been included here, but you will find the details in the 'Augmentation' chapter.

KEY WORDS

■ All words in the table

Terminology	
Active safety	Any development designed to actively avoid accidents. It can be split into four general headings: handling safety, physiological safety, perceptual safety and operational safety
Passive safety	Developments that protect the occupants of the vehicle in the event of an accident
ABS	Antilock braking system
ETS	Electronic traction system
ASR	Acceleration skid control
ESP	Electronic stability programme
ADS	Active damping system
Seat belt tensioners	In the event of an accident, a strong spring is released to pull the seat belts tighter. Some systems even use an explosive cartridge
Air bags	More and more fitted as standard, the air bag explodes in front of the driver and passengers to reduce the risk of serious injuries in the event of an accident

LEARNING TASK

➡ Look back at the key words. Explain each one to a friend, and/or write out a short description to keep as evidence.

2 Safety systems

Introduction

As noted in the table above, active safety systems are split into four areas:

Handling safety	As well as the general handling characteristics of the vehicle, as determined mostly by the steering and suspension, further systems have been developed. Antilock brakes prevent the wheels locking during braking, allowing steering control to continue. Electronic traction systems help to control wheel spin

	when moving off from an uneven or slippery surface. If a wheel starts to spin then the throttle can be backed off or the brakes applied to the spinning wheel. Acceleration skid control is similar to ETS, but tends to operate at higher speeds. Electronic stability programmes use ABS and ASR. The programme monitors acceleration speed and the angle of the steering wheel. The road wheels can be braked individually and engine power reduced to prevent loss of stability. Active damping systems control the firmness of the suspension depending on road conditions
Physiological safety	This is to keep the driver fit and alert at all times. The seat structure, heater or air conditioning operation and noise insulation systems come into this category
Perceptual safety	This means being able to see and be seen. Wiper operation is important, as is correct operation of the vehicle's lights. Heated mirrors with electric adjustment, suitable control of rain water and the windscreen reflection are all examples under this heading
Operational safety	This simply means that all the controls are in the best places and are easy to use, symbols are easy to understand and even their colours, which denote importance, are correct

Passive safety is described as developments that protect the occupants in the event of an accident. For example:

Body construction	What happens to the construction of the vehicle in the event of an accident is very important. Side impact bars and passenger cell rigidity are key issues
Interior construction	Padding and laminated construction of some components allows the interior of the vehicle to offer more protection than has been the case in the past
Restraint systems	This term covers air bags and seat belt tensioners. These systems are being fitted to more lower range vehicle all the time

KEY WORDS

- Air bag
- Energy absorbing
- Body construction
- Active and passive safety systems

Most of the systems above are discussed further in this chapter or in other parts of the book.

Seat belt tensioners

The three point fixing belt is now standard in all light vehicles, and rightly so. A further development, however, is the tensioner, which pulls the belt tighter in the event of a serious impact. An inertia mechanism is used to release a heavy-duty pre-tensioned spring, which tensions the belt within 25 milliseconds. The mechanism used by one type of seat belt tensioner is shown as Figure 11.1. Also shown in this figure are some of the general safety systems discussed above.

Body construction

The construction of the vehicle body is such that, in the event of an accident, the passengers will be protected as much as possible. Energy absorbing areas are designed-in so that the passengers remain safe inside the 'survival area'. In 75% of all frontal impacts the energy is concentrated in less than half of the front-end area. This means additional support can be designed into this section of the vehicle.

Even simple things like adjustable headrests and energy absorbing bumper supports can significantly reduce the risk of serious injury. Side impact bars are now fitted to many vehicles. Figure 11.2 shows some of these systems on a passenger car.

Advanced braking systems, air bags, programmed progressive collapse of chassis, body parts and seat belt fixings add up to the very best protection

Seat belt pretensioners react within 25 milliseconds of an impact, drawing the seatbelt tighter to the body to maximise effectiveness

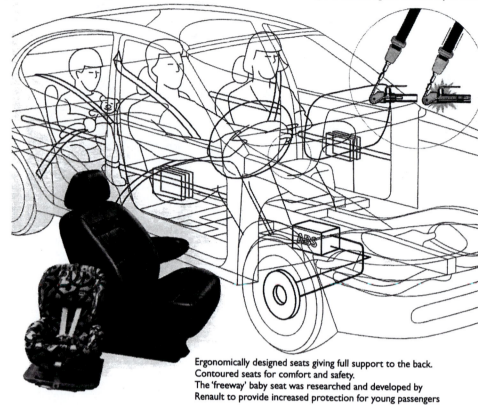

Ergonomically designed seats giving full support to the back.
Contoured seats for comfort and safety.
The 'freeway' baby seat was researched and developed by Renault to provide increased protection for young passengers

Figure 11.1 Some of the safety features employed by Renault

Air bags   A seat belt tensioner and an air bag are at present the most effective restraint system in the event of a serious accident. At speeds in excess of 40 km/h, the seat belt alone is no longer adequate. Research has determined that in 68% of cases an air bag provides a significant improvement. It is suggested that if all cars in the world were fitted with an air bag, the annual number of fatalities would be reduced by well over 50 000.

The method becoming most popular for an air bag system is to build most of the components into one unit. This reduces the amount of wiring and connections, thus improving reliability. An important aspect is that some form of system monitoring must be built in, as the operation cannot be tested – it only ever works once!

Figure 11.2 Vehicle body safety features

The sequence of events in the case of a frontal impact at about 35 km/h is detailed in the table below. Figure 11.3 also shows this sequence.

Driver in normal seating position prior to impact	About 15 ms after the impact, the vehicle is strongly decelerated and the threshold for triggering the air bag is reached
The igniter ignites the fuel tablets in the inflater	After about 30 ms, the air bag unfolds and the driver will have moved forwards as the vehicle's crumple zones collapse. The seat belt will have locked or been tensioned, depending on the system
Fuel tablets are almost spent	At 40 ms after impact, the air bag will be fully inflated and the driver's momentum will be absorbed by the air bag
Start thinking of the insurance premiums	At about 120 ms, the driver will be moved back into the seat and the air bag will have almost deflated through the side vents allowing driver visibility

Passenger air bag events are similar to the above description.

Mounting all the components in the steering wheel centre is becoming the most popular system. The main components of an air bag system are as follows:

- driver and passenger air bags
- warning light
- passenger seat switches
- pyrotechnic inflater
- igniter
- crash sensor(s)
- electronic control unit.

Figure 11.4 shows a diagram to represent the circuit of an air bag and seat belt tensioning system. The air bag is made of a nylon fabric with a coating on the inside. Prior to inflation, the air bag is folded up under suitable padding which has specially designed break lines built in. Holes are provided in the side of the air bag to allow rapid deflation after deployment. The driver's air bag has a volume of about 60 litres, and the passenger air bag about 160 litres. A warning light is used as part of the system monitoring circuit. This gives an indication of a potential malfunction, and is an important part of the circuit. Some manufacturers use a seat switch on the passenger side to prevent deployment when not occupied.

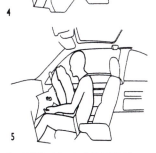

Figure 11.3 Air bag in action

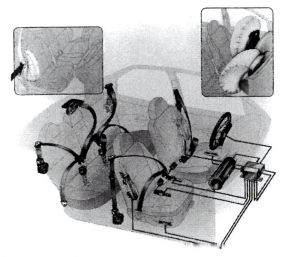

Figure 11.4 Seat belt and air bag operation

The pyrotechnic inflater and the igniter can be considered together. In the case of the driver, the inflater is located in the centre of the steering wheel. It contains a number of fuel tablets in a combustion chamber. The igniter consists of charged capacitors, which produce the ignition spark. The fuel tablets burn very rapidly and produce a quantity of nitrogen gas at a set pressure. This gas is forced into the air bag through a filter and the bag inflates, breaking through the padding in the wheel centre. After deployment, a small amount of sodium hydroxide will be present in the air bag and vehicle interior. Personal protective equipment must be used when removing the old system and cleaning the vehicle interior.

The crash sensor can take one of two forms: mechanical or electronic. The mechanical system works by a spring holding a roller in a set position. When an impact above a predetermined limit provides enough force to overcome the spring, the roller moves, triggering a switch. The switch is normally open with a resistor in parallel to allow the system to be monitored.

The other main type of crash sensor can be described as an accelerometer. This will actually sense deceleration, which is 'negative acceleration'. A severe change in the speed of the vehicle will cause an output from these sensors. Suitable electronic circuits can monitor this and be pre-programmed to react further when a signal beyond a set threshold is reached. The advantage of this technique is that the sensors do not have to be designed for specific vehicles.

The final component to be considered is the electronic control unit. In theory, when a mechanical type crash sensor is used, no electronic unit would be required. A simple circuit could be used to deploy the air bag when the sensor switch operated. However, it is the system monitoring or diagnostic part of the ECU which is most important. If a failure is detected in any part of the circuit, the warning light will be operated. Faults can be stored in the ECU memory, which can be accessed by blink code or serial fault code readers. Conventional testing of the system with a multimeter and jump wires is not recommended; it could cause the air bag to deploy.

A system using electronic sensors has about 10 ms at a vehicle speed of 50 km/h to decide if the restraint systems should be activated. In this time about 10 000 computing operations are necessary. Data for the development of these algorithms is based on computer simulations. Digital systems can also remember the events during a crash, allowing real data to be collected. Air bag systems are a major contribution to vehicle safety. Techniques vary slightly between manufacturers and I must stress that their recommendations should be followed at all times. Figure 11.5 shows a vehicle fitted with side air bags, which are currently used by some manufacturers.

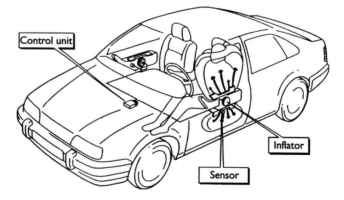

Figure 11.5 Side air bag system

3 Vehicle control systems

Anti-lock brakes
The reason for the development of anti-lock brakes is very simple. Under braking conditions, if one or more of the vehicle wheels locks or begins to skid, then this has a number of consequences:

■ braking distance increases

■ steering control is lost

■ abnormal tyre wear.

The obvious result is that an accident is more likely to occur. The maximum deceleration of a vehicle is achieved when maximum energy conversion is taking place in the brake system. This is the conversion of kinetic energy to heat energy at the discs and brake drums. The conversion process between a tyre skidding, even on a dry road, is far less than between the disks and pads. A good driver can pump the brakes on and off to prevent locking but electronic control can achieve even better results.

It is important to remember, however, that for normal use, the system is not intended to allow faster driving and shorter braking distances. It should be viewed as operating in an emergency only. Figure 11.6 shows how steering ability can be maintained even under very heavy braking conditions.

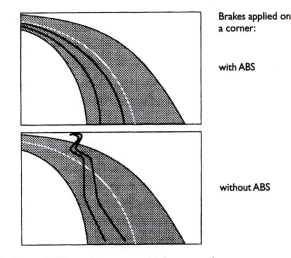

Brakes applied on a corner:

with ABS

without ABS

Figure 11.6 How ABS can improve vehicle control

Often a good way of considering a complicated system is to ask: What must the system be able to do? In other words, what are the requirements for the system?

The friction between the tyre and the road surface is a key issue when considering anti-lock brakes. Frictional forces must be transferred between the tyre contact patch and the road surface when the vehicle is accelerating or braking. The normal rules for friction between solid bodies have to be

Fail safe system	In the event of the ABS system failing, then conventional brakes must still operate to their full potential. In addition, a warning must be given to the driver. This is normally in the form of a simple warning light
Manoeuvrability must be maintained	Good steering and road holding must continue when the ABS system is operating. This is arguably the key issue, as being able to swerve round a hazard whilst still braking hard is often the best course of action
Immediate response must be available	Even over a short distance, the system must react in such a way as to make use of the best grip on the road. The response must be appropriate whether the driver applies the brakes gently or slams them on hard
Operational influences	Normal driving and manoeuvring should produce no reaction on the brake pedal. The stability and steering must be retained under all road conditions. The system must also adapt to braking hysteresis – when the brakes are applied, released and then re-applied. Even if the wheels on one side are on dry tarmac and the other side on ice, the yaw (rotation about the vertical axis of the vehicle) of the vehicle must be kept to a minimum, and only increase slowly to allow the driver to compensate
Controlled wheels	In its basic form, at least one wheel on each side of the vehicle should be controlled on a separate circuit. It is now generally common for all four wheels to be controlled separately on passenger vehicles
Speed range of operation	The system must operate under all speed conditions down to walking pace. At this very slow speed, even when the wheels lock, the vehicle will come to rest very quickly. In theory, if the wheels did not lock, the vehicle would never stop!
Other operating conditions	The system must be able to recognise aquaplaning and react accordingly. It must also still operate on an uneven road surface. The one area still not perfected is braking from slow speed on snow. The ABS will actually increase stopping distance in snow, but steering will be maintained. This is considered to be a suitable trade off

KEY WORDS

- Brake pressure
- Adaptive
- Modulator
- Control phases
- Traction
- Radar

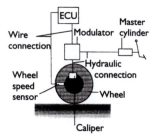

Figure 11.7 Antilock brake system

adapted because of the springy nature of rubber tyres. To get round this complicated problem, which involves a molecular theory, the term 'slip' is used to describe the action of tyre and road. Slip occurs when braking effort is applied to a rotating wheel.

The braking force depends on a number of factors; the main ones are listed:

- road surface material/condition
- tyre material, inflation pressure, tread depth, tread pattern and construction
- contact weight.

As with other systems, ABS can be considered as a central control unit with a series of inputs and outputs. The block diagram shown as Figure 11.7 represents an ABS system. The most important of the inputs are the wheel speed sensors and the main output is some form of brake system pressure control.

The task of the control unit is to compare signals from each wheel sensor, and to measure the acceleration or deceleration of an individual wheel. From this data and pre-programmed look-up tables, brake pressure to one or more of the wheels is regulated. Brake pressure can be:

- allowed to increase (Figure 11.8a)
- held constant (Figure 11.8b)
- reduced (Figure 11.8c).

All of this will also depend on the pressure on the brake pedal. A number of variables are sensed, used or controlled by the system:

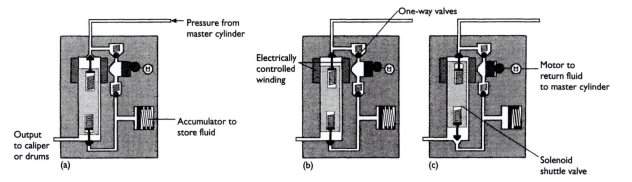

Figure 11.8 ABS modulator **a** normal pressure build up; **b** holding phase; **c** reducing

Pedal pressure	This is determined by the driver
Brake pressure	Under normal braking, this is proportional to pedal pressure but, under the control of the ABS, can be reduced, regulated or allowed to increase
Controlled variable	This is the actual result of changes in brake pressure, in other words the wheel speed, which then allows acceleration, deceleration and slip to be determined
Road/vehicle conditions	Disturbances such as the vehicle load, the state of the road, tyre condition and brake system condition

From the wheel speed sensors, the ECU calculates the following:

Vehicle reference speed	Determined by the combination of two diagonal wheels. After the start of braking, the ECU uses this value as its reference
Wheel acceleration or deceleration	This is a live measurement, which is constantly changing
Brake slip	Although this cannot be measured directly, a value can be calculated from the vehicle reference speed. This figure is then used to determine when/if ABS should take control of the brake pressure
Vehicle deceleration	During brake pressure control, the ECU uses the vehicle reference speed as the starting point and decreases it in a linear manner. The rate of decrease is determined by the evaluation of all signals received from the wheel sensors

The requirements of the anti-lock brake system can be summarised as follows:

- Rapid brake pressure reduction during wheel speed instability, so the wheel will re-accelerate fast without too much pressure reduction to avoid under-braking.
- Rapid rise in brake pressure during and after a re-acceleration to a value just less than the instability pressure.
- Discreet increase in brake pressure in the event of increased adhesion.
- Sensitivity suited to the prevalent conditions.
- Anti-lock braking must not be initiated during axle vibration.

ABS operation:

- Initial braking, ABS not yet activated.
- Wheel speed exceeds a threshold calculated from the vehicle reference speed and brake pressure is held at a constant value.
- Wheel deceleration falls below a threshold and brake pressure is reduced.
- Brake pressure holding is now occurring and wheel speed will increase.

- Wheel acceleration exceeds the upper limit so brake pressure is now allowed to increase.
- Pressure is again held constant as the limit is exceeded.
- Brake pressure is now increased in stages until wheel speed threshold is exceeded.
- Brake pressure is decreased again and then held constant.

The above process continues until the brake pedal is released or the vehicle speed is less than a set minimum, at which time the wheels will lock to finally bring the vehicle to rest

There are a few variations between manufacturers as to how the ABS actually operates. This involves a number of different components. The majority of systems consist of just three main components: the wheel speed sensors, the electronic control unit and a hydraulic modulator.

Wheel speed sensors	These devices are inductive sensors and work in conjunction with a toothed wheel. They consist of a permanent magnet and a soft iron rod around which is wound a coil of wire. As the toothed wheel rotates, the changes in inductance of the magnetic circuit generate a signal with a frequency that is proportional to wheel speed
Electronic control unit	The function of the ECU is to take in information from the wheel sensors and calculate the best course of action for the hydraulic modulator. The heart of a modern ECU consists of two microprocessors, which run the same programme independently of each other. This ensures greater security against any fault which could affect braking performance. If a fault is detected, the ABS disconnects itself and operates the warning light
Hydraulic modulator	This device has three operating positions as shown in the diagram (Figure 11.8). These are: ■ pressure release ■ pressure holding ■ pressure build-up. The position of these valves is controlled by electrical windings. The pump motor also runs when ABS is activated.

Traction control The steerability of a vehicle is not only lost when the wheels lock up on braking. The same effect arises if the wheels spin when driving off under hard acceleration. Electronic traction control has been developed as a supplement to ABS. This control system prevents the wheels from spinning when moving off or when accelerating sharply while on the move. An individual wheel which is spinning can be braked in a controlled manner. If both or all of the wheels are spinning, the drive torque is reduced by means of engine control. Traction control has become known as ASR, ETS or TCR!

Traction control is not available as an independent system, but in combination with ABS. This is because the majority of components are the same as for ABS. Traction control requires a change in control in the ECU and a few extra control actuators, such as to control the throttle opening. Figure 11.9 shows a block diagram of a traction control system.

Traction control will intervene to achieve the following:

- maintain stability
- reduction of yawing moment reactions
- provide optimum propulsion at all speeds
- reduce driver workload.

An automatic control system can intervene in many cases more quickly and precisely than the driver of the vehicle. This allows stability to be maintained at a time when the driver might not have been able to cope with the situation.

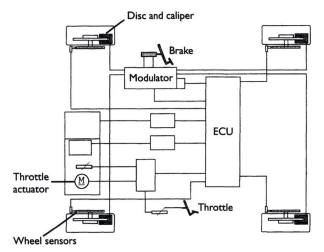

Figure 11.9 Traction control system

Control of tractive (driving) force is achieved by the following methods:

Throttle control	An actuator moves the throttle cable, or if the vehicle employs a drive by wire accelerator, then control will be in conjunction with the engine management. The throttle control is independent of the driver's throttle pedal position
Ignition control	If ignition timing is retarded, the engine torque can be reduced by up to 50% in a very short space of time
Braking effect	If the spinning wheel is restricted by brake pressure, the reduction in torque at the effected wheel is very fast

The layout of a traction control system, which includes links with other vehicle control systems, is shown as Figure 11.9.

Cruise control Cruise control is the ideal example of a closed loop control system. Figure 11.10 illustrates this in the form of a block diagram. The purpose of cruise control is to allow the driver to set the vehicle speed and, when the cruise control is activated, the speed of the vehicle is maintained automatically. The system must react to the measured speed of the vehicle and adjust the throttle accordingly. The reaction time of the system is important so that the vehicle speed does not feel to be surging up and down. Figure 11.11 shows the components of a cruise control system.

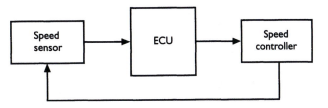

Figure 11.10 Cruise control

Other facilities are often included, such as allowing the speed to be gradually increased or decreased at the touch of a button. Most systems also remember the last set speed and will resume to this again at the touch of a button. The following is the list of requirements for a good cruise control system:

■ hold the vehicle speed at the selected value

■ hold the speed with minimum surging

■ allow the vehicle to change speed

1. Cruise ECU
2. Cruise mainswitch
3. Set button
4. Resume button
5. Road speed transducer
6. Brake switch
7. Clutch switch
8. Shift lever switch (automat
9. Cruise actuator
10. Stepper motor cable

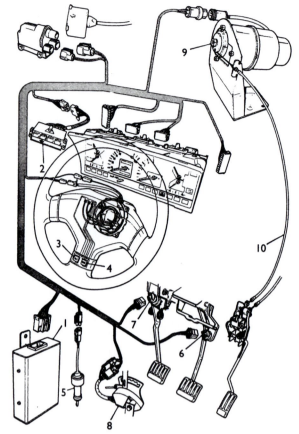

Figure 11.11 Cruise control system with a motor drive

■ relinquish control immediately if either the brakes or clutch is applied

■ store the last set speed

■ contain built-in safety features.

A main switch switches on the cruise control, which, in turn, is ignition controlled. Most systems do not retain the speed setting in memory when the main switch has been turned off. Operating the 'set' switch programs the memory but this will normally only work if conditions similar to the following are met:

■ vehicle speed is greater than 40 km/h

■ vehicle speed is less than 200 km/h

■ change of speed is less than 8 km/h/s

■ automatics must be in 'drive'

■ brakes or clutch are not being operated

■ engine speed is stable.

Once the system is set, the speed is maintained to within about 3–4 km/h until it is deactivated by pressing the brake or clutch pedal, pressing the 'resume' switch or turning off the main control switch. Last speed is retained in memory except when the main switch is turned off. If the cruise control system is required again, then either the 'set' button will hold the vehicle at its current speed, or the 'resume' button will accelerate the vehicle to the previous set speed. When cruising at a set speed, if the driver presses and holds the 'set' button, the vehicle will accelerate until the desired speed is reached and the button is released. If the driver accelerates from the set speed to overtake, for example, then when the throttle is released, the vehicle will slow down until it reaches the last set position. The main components of a typical cruise control system are as follows:

Actuator *[Diagram: To vent, To inlet manifold via a check valve, Vent valve y, Safety valve z, Control winding, Vacuum valve x, Diaphragm, Throttle linkage]* **Figure 11.12** Cruise control vacuum actuator	A number of methods are used to control the throttle position. Vehicles fitted with drive by wire systems allow the cruise control to operate the same actuator. A motor can be used to move the throttle cable, or a vacuum operated diaphragm with control valves. This technique is shown as Figure 11.12. When the speed needs to be increased, valve 'x' is opened, allowing low pressure from the inlet manifold to one side of the diaphragm. The atmospheric pressure on the other side will move the diaphragm and, hence, the throttle. To move the other way, valve 'x' is closed and valve 'y' is opened, allowing atmospheric pressure to enter the chamber. The spring moves the diaphragm back. If both valves are closed, then the throttle position is held. Valve 'x' is normally closed and valve 'y' normally open, thus, in the event of electrical failure, cruise will not remain engaged and the manifold vacuum is not disturbed. Valve 'z' provides extra safety and is controlled by the brake and clutch pedals
Main switch and warning lamp	This is a simple on/off switch that should be located in easy reach of the driver. The warning lamp can be part of this switch, or part of the main instrument display, as long as it is in the driver's field of vision
Set and resume switches	These are fitted either on the steering wheel or on a stalk from the steering column. When part of the steering wheel, slip rings are needed to transfer the electrical connection. The 'set' button programmes the speed into memory, and can also be used to increase the vehicle and memory speed. The 'resume' button allows the vehicle to reach its last set speed
Brake switch	This switch is very important; it would be dangerous braking if the cruise control system was trying to maintain the vehicle speed. This switch is normally of high quality and is fitted in place of or as a supplement to, the brake light switch. The brake pedal activates it
Clutch or automatic gearbox switch	The clutch switch is fitted in a similar way to the brake switch. It deactivates the cruise system to prevent the engine speed increasing if the clutch is pressed. The automatic gearbox switch will only allow the cruise to be engaged when in the 'drive' position. This is to prevent the engine over speeding if the cruise control tried to accelerate to a high road speed with the gear selector in position '1' or '2'
Speed sensor	This will often be the same sensor that is used for the speedometer. A pulsed signal is produced, with a frequency proportional to the vehicle speed

Conventional cruise control has now developed to a high degree of quality. It is, however, not always very practical on many European roads as the speed of the general traffic is constantly varying and often very heavy. The driver has to take over from the cruise control system on many occasions to speed up or slow down. Adaptive cruise control can automatically adjust the vehicle speed to the current traffic situation. The system has three main aims:

■ maintain a speed as set by the driver
■ adapt this speed and maintain a safe distance from the vehicles in front
■ provide a warning if there is a risk of collision.

The main components of an adaptive cruise system are shown in Figure 11.13. The key extra component is the headway sensor, which is a form of radar. It contains a transmitter and receiver unit.

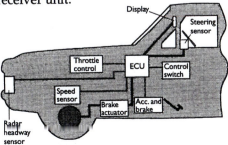

Figure 11.13 Adaptive cruise control

The operation of an adaptive cruise system is the same as a conventional system but, when a signal from the headway sensor detects an obstruction, the vehicle speed is decreased. If the optimum stopping distance cannot be achieved by just backing off the throttle, a warning is supplied to the driver. More complex systems could also take control of the vehicle transmission and brakes. This is a little worrying! It is important to note that adaptive cruise control is designed to relieve the burden on the driver, not take full control of the vehicle.

Obstacle avoidance radar

This system, sometimes called collision avoidance radar, is most commonly used as an aid to reversing. It is used to give the driver some indication as to how much space is behind the car.

The principle of radar as a reversing aid is illustrated by Figure 11.14. The output to the driver can be visual or audible, the latter being most appropriate, as the driver is likely to be looking backwards. The audible signal is a 'pip-pip-pip' type sound, the repetition frequency of which increases as the car comes nearer to the obstruction. The signal becomes almost continuous when impact is imminent!

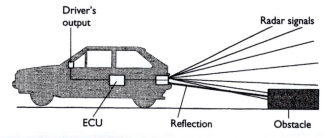

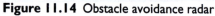

Figure 11.14 Obstacle avoidance radar

The operation of a basic radar system is as follows:

- a radio transmitter generates radio waves, which are then radiated from an antenna
- this causes 'lighting up' of the airspace with radio waves
- a target, such as a vehicle in this space, scatters a small portion of the radio energy back to a receiving antenna
- this weak signal is amplified
- the distance (range) is determined because radio waves travel at a known speed (3×10^8 m/s)
- the driver's display is operated, or the warning noise activated.

LEARNING TASKS

➡ Look back at the key words. Explain each one to a friend, and/or write out a short description to keep as evidence.

➡ Make a simple sketch to show an ABS block diagram.

➡ Examine a real system and note the traction control and ABS components.

---

4 Heating and air conditioning

Introduction

Any heating and ventilation system has a simple set of functional requirements. These can be summarised as follows:

- adjustable temperature in the vehicle cabin
- heat must be available as soon as possible
- distribute heat to various parts of the vehicle
- ventilate with fresh air with minimum noise
- demisting of all windows
- ease of control operation.

Some solutions to these requirements are discussed below, starting with simple ventilation and leading on to automatic temperature control.

Ventilation and heating – the water cooled engine

KEY WORDS

- Refrigerant
- Condenser
- Evaporator
- Heat
- Temperature
- Screen heating

To allow fresh air from outside the vehicle to be circulated inside the cabin, a pressure difference must be created. This is achieved by using a plenum chamber. A plenum chamber, by definition, holds a gas (in this case air) at a pressure higher than the ambient pressure. The plenum chamber on a vehicle is usually situated just below the windscreen, behind the bonnet. When the vehicle is moving, the airflow over the vehicle will cause a higher pressure in this area. Suitable flaps and drains are utilised to prevent water entering the car through this opening.

By means of distribution trunking, control flaps and suitable 'nozzles', the air can now be directed as required. This system is enhanced with the addition of a variable speed blower motor. Figure 11.15 shows a ventilation and heating system layout. When extra air is forced into a vehicle cabin, the interior pressure would increase if no outlet was available. Most passenger cars have the outlet grills on each side of the vehicle above the rear quarter panel.

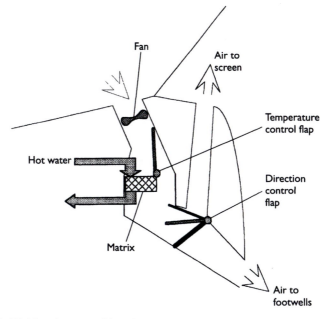

Figure 11.15 Ventilation and heating system

Heat from the engine is utilised to increase the temperature of the car interior. Using a heat exchanger, often called a heater matrix, does this. Due to the action of the thermostat in the engine cooling system, the water temperature remains nearly constant. This means the air being passed over the heater matrix is heated by a set amount.

Some form of control is required over how much heat is required. The method used on most modern vehicles is the blending technique. This is simply a control flap, which determines how much of the air being passed into the vehicle is directed over the heater matrix.

By a suitable arrangement of flaps, it is possible to direct air of the chosen temperature to selected areas of the vehicle interior. In general, basic systems allow the warm air to be adjusted between the inside of the windscreen and the driver and passenger footwells. Most vehicles also have small vents directing warm air at the driver's and front passenger's side windows. Fresh, cool air outlets with directional nozzles are also fitted.

One more facility available on most vehicles is the choice between fresh or recirculated air. The reason for this is to decrease the time taken to demist or defrost the vehicle windows and to heat the car interior more quickly. Another reason is that the outside air may not be very clean. A multiposition switch controls the speed of the motor by using different resistors in the circuit. In most cases, the motor is controlled to three or four set speeds.

Air conditioning

A vehicle fitted with air conditioning allows the temperature of the cabin to be controlled to the ideal or most comfortable value determined by the ambient conditions. Air conditioning can be manually controlled or, as is often the case, combined with some form of electronic control. The system as a whole can be thought of as a type of refrigerator or heat exchanger. Heat is removed from the car interior and dispersed to the outside air.

To understand the principle of refrigeration, the following terms and definitions will be useful:

- heat is a form of energy
- temperature is the degree or intensity of heat of a body, and the condition that determines whether or not it will transfer heat to, or receive heat from, another body
- heat will only flow from a higher to a lower temperature
- change of state is a term used to describe the changing of a solid to a liquid, a liquid to a gas, a gas to a liquid or a liquid to a solid
- evaporation is used to describe the change of state from a liquid to a gas
- condensation is used to describe the change of state from gas to liquid
- latent heat describes the energy required to evaporate a liquid without changing its temperature.

Latent heat in the change of state of a refrigerant is the key to air conditioning. As an example of this, put a liquid, such as methylated spirits, on your hand – it feels cold. This is because it evaporates and the change of state (liquid to gas) uses heat from your body. This is why the process is often thought of as 'unheating' rather than cooling. The refrigerant used in many air conditioning systems changes state from liquid to gas at $-26.3°C$.

Figure 11.16 shows the layout of an air conditioning or refrigeration system. The main components are the evaporator, the condenser and the pump or compressor. The evaporator is situated in the car, the condenser outside the car, usually in the air stream, and the compressor is driven by the engine.

As the pump operates, it causes the pressure on its intake side to fall. This allows the refrigerant in the evaporator to evaporate and draw heat from the vehicle interior. The high pressure or output of the pump is connected to the condenser. The pressure causes the refrigerant to condense (in the condenser), giving off heat outside the vehicle as it changes state.

The compressor pumps low pressure but heat laden vapour from the evaporator, compresses it and pumps it as a super heated vapour under high pressure to the condenser. The temperature of the refrigerant at this stage is much higher than the outside air temperature, hence it gives up its heat via the fins on the condenser as it changes state back to a liquid. This high-

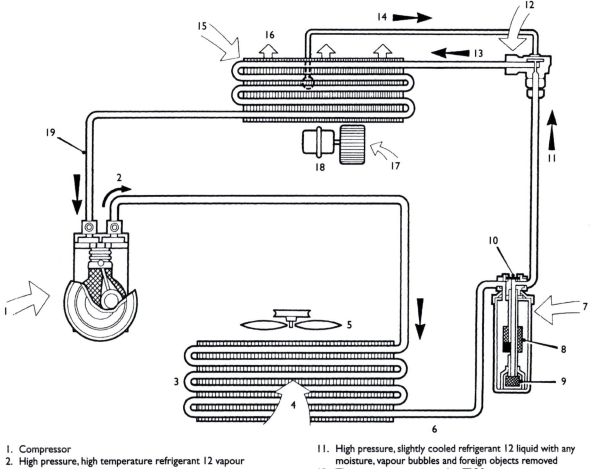

1. Compressor
2. High pressure, high temperature refrigerant 12 vapour
3. Condenser
4. Air flow through condenser
5. Cooling fan to boost air flow through condenser
6. High pressure, slightly cooled refrigerant 12 liquid
7. Receiver-drier
8. Drying agent
9. Filter
10. Sight glass

11. High pressure, slightly cooled refrigerant 12 liquid with any moisture, vapour bubbles and foreign objects removed
12. Thermostatic expansion valve (TXV)
13. Low pressure, low temperature atomised liquid refrigerant 12
14. Capillary tube with thermal bulb fitted to evaporator outlet pipe, this is to monitor temperature, thus controlling the refrigerant 12 flow
15. Evaporator
16. Air flow into car
17. Sirocco fan
18. Fan motor
19. Low pressure, slightly warmed refrigerant 12 vapour

Figure 11.16 Air conditioning system layout

pressure liquid is then passed to the receiver drier to store any vapour which has not yet turned back to a liquid. A dessicant (drying agent) bag removes any moisture (water) that is contaminating the refrigerant. The high-pressure liquid is now passed through the thermostatic expansion valve and is converted back to a low-pressure liquid as it passes through a restriction in the valve into the evaporator. As the liquid changes state to a gas in the evaporator, it takes up heat from its surroundings, thus cooling or 'unheating' the air which is forced over the fins. The low-pressure vapour leaves the evaporator returning to the pump, thus completing the cycle. If the temperature of the refrigerant increases beyond certain limits, the condenser cooling fans can be switched in to supplement the ram air effect.

Figure 11.17 shows an air conditioning system compressor. It is belt driven from the engine crankshaft and it causes refrigerant to circulate through the system. The compressor is controlled by an electromagnetic clutch, which may be either under manual control or electronic control depending on the type of system.

Figure 11.18 shows a condenser fitted in front of the vehicle radiator. It is very similar in construction to the radiator and fulfils a similar role. The heat is conducted through the aluminium pipes and fins to the surrounding air and then, by a process of radiation and convection, is dispersed by the air movement.

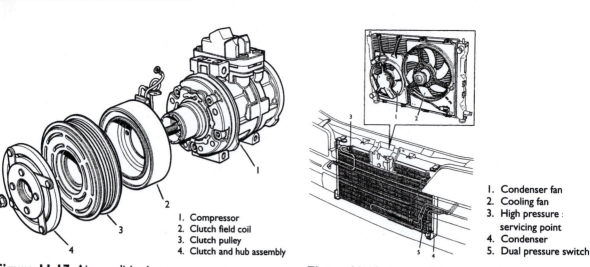

1. Compressor
2. Clutch field coil
3. Clutch pulley
4. Clutch and hub assembly

Figure 11.17 Air conditioning compressor

1. Condenser fan
2. Cooling fan
3. High pressure servicing point
4. Condenser
5. Dual pressure switch

Figure 11.18 Air conditioning condenser

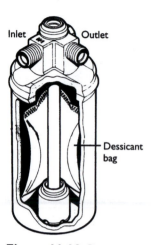

Inlet   Outlet

Dessicant bag

Figure 11.19 Receiver drier

Figure 11.19 shows a typical receiver drier assembly. It is connected in the high-pressure line between the condenser and the thermostatic expansion valve. This component carries out four tasks:

- a reservoir to hold refrigerant until a greater flow is required
- a filter to prevent contaminants circulating through the system
- vapour is retained in this unit until it finally converts back to a liquid
- a drying agent removes any moisture from the system.

A sight glass is fitted to some receiver driers to give an indication of refrigerant condition and system operation. The refrigerant generally appears clear if all is in order.

The thermostatic expansion valve is shown in Figures 11.16 and 11.20. The main function of the valve is to control the flow of refrigerant as demanded by the system. This, in turn, controls the temperature of the evaporator.

Figure 11.20 shows a typical evaporator assembly. It is similar in construction to the condenser, consisting of fins to maximise heat transfer.

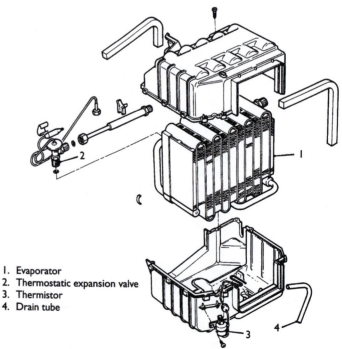

1. Evaporator
2. Thermostatic expansion valve
3. Thermistor
4. Drain tube

Figure 11.20 Evaporator

It is mounted in the car under the dash panel forming part of the overall heating and ventilation system. As well as cooling the air passed over it, the evaporator also removes moisture from the air. This is because the moisture in the air condenses on the fins and is drained away. The action is much like breathing on a cold pane of glass.

Automatic temperature control system

Full temperature control systems provide a comfortable interior temperature in line with the driver's requirements. The electronic control unit has control of:

- fan speed
- air distribution
- air temperature flaps
- fresh or recirculated air flaps
- air conditioning pump.

A number of sensors can be used to provide input to the ECU.

- ambient temperature sensor mounted outside the vehicle to allow compensation for extreme temperature variations. This device is usually a thermistor.
- solar light sensor mounted on the fascia panel.
- the car temperature sensor is a thermistor but, to allow for an accurate reading, a small motor and fan can be used to take a sample of interior air and direct it over the sensor.
- a coolant temperature sensor is used to monitor the temperature of the coolant supplied to the heater matrix.

The ECU takes information from all of the above sources and can set the system in the most appropriate manner. Control of the flaps can be either by solenoid controlled vacuum actuators or by small motors. The main blower motor is often controlled by a heavy-duty power transistor, which gives a constantly variable speed. These systems can provide a comfortable interior temperature in exterior conditions ranging from −10 to +35°C, and in bright sunlight.

Front and rear screen heating

Heating of the rear screen involves a circuit with a relay and usually a timer. The heating elements are thin metallic strips bonded to the glass. When a current is passed through the elements, heat is generated and the window will defrost or demist. This circuit can draw high current, 10 A to 15 A being typical. Because of this, the circuit now often contains a timer relay to prevent the heater being left on too long. The timer will switch off after about 10 to 15 minutes. The elements are usually shaped to defrost in the rest position of the rear wiper blade, if fitted.

Front windscreen heating is used on some top-of-the-range vehicles. This, of course, presents more problems than the rear screen, as vision must not be obscured. The technology is drawn from the aircraft industry and involves very thin wires cast into the glass. As with the heated rear window, this device can consume a large current and, again, uses a timer relay.

Seat heaters

The concept of seat heating is simple. A heating element is placed in the seat, together with an on-off switch and a control to regulate the heat. However, the design of these heaters is more complex than first appears.

The heater must meet the following criteria:

- the heater must only supply the heat loss experienced by the person's body

- heat is only to be supplied at the major contact points
- leather and fabric seats require different systems due to their different thermal properties
- heating elements must fit the design of the seat
- the elements must pass the same rigorous tests as the seat, such as squirm, jounce and bump tests.

The traditional method of control is a simple thermostat switch. Recent developments, however, tend to favour electronic control combined with a thermistor. These seat heaters will heat up to provide an initial sensation in one minute and to full-regulated temperature in three minutes. Very nice on a cold morning!

LEARNING TASKS

- ➡ Look back at the key words. Explain each one to a friend, and/or write out a short description to keep as evidence.
- ➡ Make a simple sketch to show a screen heater circuit.
- ➡ Write a short explanation about how air conditioning works. Why does it also demist the windscreen?

5 Vehicle body systems

Electric seats, mirrors and sun roof operation

KEY WORDS

- Reverse circuit
- CDL
- Change-over relay
- One shot
- Back off

Electrical movement of seats, mirrors and the sunroof are included in this one section as the operation of each system is quite similar. All the above mentioned systems operate using one or several permanent magnet motors. A supply reversing circuit is also used. A typical motor reverse circuit is shown as Figure 11.21. When the switch is moved, one of the relays will

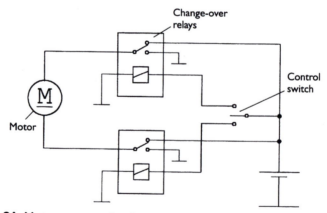

Figure 11.21 Motor reverse circuit

operate and change the polarity of the supply to one side of the motor. If the switch is moved the other way, the polarity of the other side of the motor is changed. When at rest, both sides of the motor are at the same potential.

Adjustment of a seat is done by using a number of motors to allow positioning of different parts. Movement is possible in the following ways:

- front to rear
- cushion height rear
- cushion height front

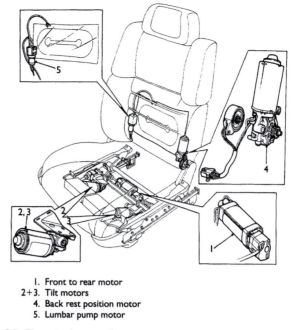

1. Front to rear motor
2+3. Tilt motors
4. Back rest position motor
5. Lumbar pump motor

Figure 11.22 Electrical seat adjustment components

■   backrest tilt

■   headrest height

■   lumber support.

Figure 11.22 shows a typical electrically controlled seat. This system uses four positioning motors and one smaller motor to operate a pump, which controls the lumber support bag. Each motor can be considered to operate by a simple rocker type switch, which controls two relays as described above. Nine relays are required altogether! Two for each motor and one to control the main supply. When seat position is set, some vehicles have set position memories to allow automatic re-positioning if the seat has been moved.

Many vehicles now have electrical adjustment of mirrors, particularly on the passenger side. The system used is much the same as has been discussed above in relation to seat movement. Two small motors are used to move the mirror vertically and horizontally. Many mirrors now also contain a small heating element on the rear of the glass. This is operated for a few minutes when the ignition is first switched on, and can also be linked to the heated rear window circuit.

The operation of an electric sunroof is similar to the motor reverse circuit shown as Figure 11.21. However, further components and circuitry are needed to allow the roof to slide, tilt and stop in the closed position. The extra components used are a micro switch and a latching relay. A latching relay works in much the same way as a normal relay except that it locks into position each time it is energised. The mechanism used to achieve this is much like that used in ballpoint pens, which use a button on top. The micro switch is mechanically positioned so as to operate when the roof is in its closed position. A rocker switch allows the driver to adjust the roof.

Central locking   When the key is turned in the driver's door lock, all the other doors on the vehicle should also lock. Motors or solenoids in each door achieve this. If the system can only be operated from the driver's door key, then an actuator is not required in this door. If the system can be operated from either front door or by remote control, then all the doors need an actuator. Vehicles with sophisticated alarm systems often lock all the doors as the alarm is set.

Many door actuators are small motors which, via suitable gear reduction, operate a rod in either direction to lock or unlock the doors. A motor reverse circuit is used to achieve the required action.

Figure 11.23 shows a door locking circuit. The main control unit contains two change over relays which are actuated by either the door lock switch or, if fitted, the remote infra-red key. The motors for each door lock are wired in parallel, and all operate at the same time.

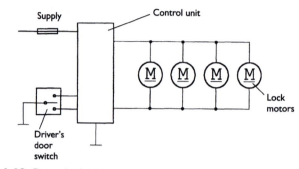

Figure 11.23 Door lock circuit

A small hand-held transmitter and a receiver unit control remote central door locking. When the infra-red key is operated by pressing a small switch, a complex code is transmitted. The receiver picks up this code and sends it in an electrical form to the main control unit. If the code is correct, the relays are triggered and the door locks are either locked or unlocked. If an incorrect code is received on three consecutive occasions when attempting to unlock the doors, the system will switch itself off until the door is opened by the key. This will also reset the system and allow the correct code to again operate the locks. This technique prevents a scanning type transmitter unit from being used to open the doors.

Electric window operation

The basic form of electric window operation is similar to many of the systems discussed above. That is, a motor reversing system either by relays or directly by a switch. More sophisticated systems are becoming more popular for safety reasons as well as improved comfort. The following features are now available from many manufacturers:

- one shot up or down
- inch up or down
- lazy lock
- back off.

The complete system consists of an electronic control unit containing the window motor relays, switch packs and a link to the door lock and sunroof circuits. This is represented in the form of a block diagram in Figure 11.24. When a window is operated in one shot or one touch mode, the window is driven in the chosen direction until either the switch position is reversed, the motor stalls or the ECU receives a signal from the door lock circuit. The problem with one shot operation is that if a child should become trapped in the window, there is a serious risk of injury. To prevent this, the back off feature is used. An extra contact ring is fitted to the motor armature and produces a signal proportional to the motor speed. If the 'rate of change of speed' of the motor is detected as being below a certain threshold when closing, the ECU will reverse the motor until the window is fully open.

By counting the number of pulses received, the ECU can also determine the window position. This is important, as the window must not reverse when it stalls in the closed position. In order for the ECU to 'know' the window position, it must be initialised. This is often done by operating the motor to drive the window first fully open, and then fully closed. If this is not done, the one shot close will not operate.

A lazy lock feature allows the car to be fully secured by one operation of a remote key. This is done by a link between the door lock ECU, and the window and sunroof ECU's. A signal is supplied and causes all the windows to close, the sunroof shuts and finally the doors lock. The alarm will also be set if required. The windows may close in turn to prevent excessive current demand.

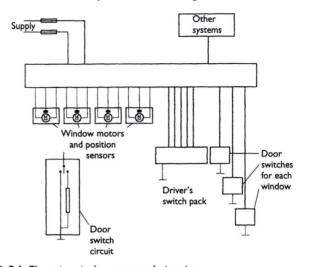

Figure 11.24 Electric window control circuit

A circuit for electric windows is shown as Figure 11.24. Note the connections to other systems such as door locking and the rear window isolation switch. This is commonly fitted to allow the driver to prevent children operating the rear windows.

LEARNING TASKS

➡ Look back at the key words. Explain each one to a friend, and/or write out a short description to keep as evidence.

➡ Examine a real system and note the component layout.

➡ Write a short explanation about the advantages of 'lazy lock' systems.

---

6 Security

Alarmed – you should be!

Stolen cars and theft from cars account for about a quarter of all reported crime. Up to 500 000 cars are reported missing each year in the UK, and about 100 000 are never recovered. Even when returned, many are damaged. Most car thieves are opportunists; even a basic alarm system serves as a deterrent.

Car and alarm manufacturers are constantly striving to improve security. Building the alarm system as an integral part of the vehicle electronics has made improvements. Even so, retrofit systems can still be very effective. Alarms are activated in a number of ways:

KEY WORDS

■ Coded ECU
■ Key code
■ Immobilise
■ Remote key

■ switches on all entry points

■ battery voltage sensing

■ volumetric sensing.

The alarms also use different ways of disabling the vehicle, as well as sounding a siren and flashing the lights:

- ignition circuit cut off
- starter circuit cut off
- engine ECU code lock.

A separate switch or a remote transmitter is used to set the alarm. The most common systems are set automatically when the doors are locked with the remote key. Most systems now come with two remote keys that use small button type batteries and have an LED that shows when the signal is being sent. When operating with flashing lights, most systems draw about 5 A. Without flashing lights (siren only), the current draw is less than 1 A. The sirens produce a sound level of about 95 dB, when measured 2 m in front of the vehicle. Figure 11.25 shows a block diagram of a complex alarm system.

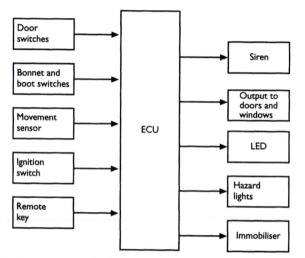

Figure 11.25 Alarm system block diagram

One idea currently in use is to program a security code into the engine electronic control unit. This can only be 'unlocked', to allow the engine to start, when it receives a coded signal. Some Ford cars use a special ignition key, which is programmed with the required information. Even the correctly 'cut' key will not start the engine. Citroen, in many top range models, use a similar idea but the code has to be entered via a numerical keypad. Of course, nothing will stop the car being lifted on to a lorry and driven away, but this technique will mean a new engine control ECU will be needed. The cost will be high and also questions may be asked as to why a new ECU is required.

Alarm systems continue to develop!

LEARNING TASKS

- Look back at the key words. Explain each one to a friend, and/or write out a short description to keep as evidence.
- Examine a real system and note the component layout.
- Read about different alarm systems. What are the advantages of one against another?
- Write a short explanation about how and why incorrect signals from a remote key cause the system to shut off.

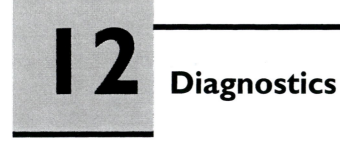

12 Diagnostics

1 Introduction

Start here! In the level 2 book, I included diagnostics or fault finding as part of each technology chapter. This time it is all included under this one heading. The reason is that now you are progressing with your studies, you will have realised that the subject of diagnostics does not necessarily relate to individual areas. In other words, once your knowledge of a vehicle system is at a suitable level, then you will use the same logical *process* of diagnosing the fault on all systems. I hope you can remember my six stages of fault diagnosis:

1. Verify the fault.
2. Collect further information.
3. Evaluate the evidence.
4. Carry out further tests in a logical sequence.
5. Rectify the problem.
6. Check all systems.

Figure 12.1 Complex systems require logical diagnostic procedures

I have included a number of diagnostic checklists and courses of action in this chapter, but they are by no means exhaustive. Let me know by visiting my Web site of any suggestions you have or examples of diagnostic routines you use. I can then make the ideas available to others. The checklists and test sequences included here are mostly related to stages 3 and 4 of the fault finding sequence above. A very important point to remember is to always work in a logical manner. Most tables are set out as follows:

Fault or symptoms	Possible cause
A brief description of the problem	Suggestions as to the cause. If these are numbered it suggests a sequence to follow

The complexity of modern vehicle systems means it is only possible to give general examples of the faults and possible causes but, with a little thought, the principles can be applied in all areas.

Terminology

Most of the terminology for this chapter is covered elsewhere. However, the following are important to understand:

Diagnostics	The process of identifying a fault by means of its symptoms
UPK	Underpinning knowledge of a system is vital if you are to diagnose faults
Logical procedure	The only way to find a fault is to work in such a way that you will not miss anything
Report	When you find a fault, you may have to report your findings before a repair is carried out

Figure 12.2 You need UPK of all vehicle systems to diagnose faults

Test equipment

Most of the test equipment you will need has been covered in the level 2 book. Some further details are, however, provided in chapter 3 of this book – in particular the use of an oscilloscope. You will notice that, although test equipment is vital to make a correct diagnosis, your *observations* are just as important.

LEARNING TASK

➡ Look back at the key words. Explain each one to a friend, and/or write out a short description to keep as evidence.

2 Vehicle systems

Engines

Please note that this section covers related engine systems as well as the engine itself.

Fault or symptom	Possible cause
Engine does not rotate when trying to start	Battery connection loose or corroded
	Battery discharged or faulty
	Broken, loose or disconnected wiring in the starter circuit
	Defective starter switch or autobox inhibitor switch
	Starter pinion or flywheel ring gear loose
	Earth strap broken, loose or corroded
Engine rotates but does not start	No fuel in the tank!
	Discharged battery (slow rotation)
	Battery terminals loose or corroded
	Air filter dirty or blocked
	Low cylinder compressions
	Broken timing belt
	Damp ignition components
	Fuel system fault
	Spark plugs worn to excess
	Ignition system open circuit
Difficult to start when cold	Discharged battery (slow rotation)
	Battery terminals loose or corroded
	Air filter dirty or blocked
	Low cylinder compressions
	Fuel system fault
	Spark plugs worn to excess
	Enrichment device not working (choke or injection circuit)
Difficult to start when hot	Discharged battery (slow rotation)
	Battery terminals loose or corroded
	Air filter dirty or blocked
	Low cylinder compressions
	Fuel system fault
Starter noisy	Starter pinion or flywheel ring gear loose
	Starter mounting bolts loose
	Starter worn (bearings etc)
	Discharged battery (starter may jump in and out)
Starter turns engine slowly	Discharged battery (slow rotation)
	Battery terminals loose or corroded
	Earth strap or starter supply loose or disconnected
	Internal starter fault
Engine starts but then stops immediately	Ignition wiring connection intermittent
	Fuel system contamination
	Fuel pump or circuit fault (relay)
	Intake system air leak
	Ballast resistor open circuit (older cars)
Erratic idle	Air filter blocked
	Incorrect plug gaps
	Inlet system air leak
	Incorrect CO setting
	Uneven or low cylinder compressions (maybe valves)
	Fuel injector fault
	Incorrect ignition timing
	Incorrect valve timing
Misfire at idle speed	Ignition coil or distributor cap tracking
	Poor cylinder compressions
	Engine breather blocked
	Inlet system air leak
	Faulty plugs

Fault or symptom	Possible cause
Misfire through all speeds	Fuel filter blocked
	Fuel pump delivery low
	Fuel tank ventilation system blocked
	Poor cylinder compressions
	Incorrect plugs or plug gaps
	HT leads breaking down
Engine stalls	Idle speed incorrect
	CO setting incorrect
	Fuel filter blocked
	Air filter blocked
	Intake air leak
	Idle control system not working
Lack of power	Fuel filter blocked
	Air filter blocked
	Ignition timing incorrect
	Low fuel pump delivery
	Uneven or low cylinder compressions (maybe valves)
	Fuel injectors blocked
	Brakes binding or clutch slipping
Backfires	Incorrect ignition timing
	Incorrect valve timing (cam belt not fitted correctly)
	Fuel system fault (air flow sensor on some cars)
Oil pressure gauge low or warning light on	Low engine oil level
	Faulty sensor or switch
	Worn engine oil pump and/or engine bearings
	Engine overheating
	Oil pickup filter blocked
	Pressure relief valve not working
Runs on when switched off	Ignition timing incorrect
	Idle speed too high
	Anti-run on device not working
	Carbon build up in engine
	Engine overheating
Pinking or knocking under load	Ignition timing incorrect
	Ignition system fault
	Carbon build up in engine
	Knock sensor not working
Sucking or whistling noises	Leaking exhaust manifold gasket
	Leaking inlet manifold gasket
	Cylinder head gasket
	Inlet air leak
	Water pump or alternator bearing
Rattling or tapping	Incorrect valve clearances
	Worn valve gear or camshaft
	Loose component
Thumping or knocking noises	Worn main bearings (deep knocking/rumbling noise)
	Worn big-end bearings (heavy knocking noise under load)
	Piston slap (worse when cold)
	Loose component
Rumbling noises	Bearings on ancillary component

Cooling

Fault or symptom	Possible cause
Overheating	Low coolant level (may be due to a leak) Thermostat stuck closed Radiator core blocked Cooling fan not operating Temperature gauge inaccurate Airlock in system (some systems have a complex bleeding procedure) Pressure cap faulty
Overcooling	Thermostat stuck open Temperature gauge inaccurate Cooling fan operating when not needed
External coolant leak	Loose or damaged hose Radiator leak Pressure cap seal faulty Water pump leak from seal or bearing Boiling due to overheating or faulty pressure cap Core plug leaking
Internal coolant leak	Cylinder head gasket leaking Cylinder head cracked
Corrosion	Incorrect coolant (antifreeze, etc.) Infrequent flushing
Freezing	Lack of antifreeze Incorrect antifreeze

Fuel

Fault or symptom	Possible cause
Excessive consumption	Blocked air filter Incorrect CO adjustment Fuel injectors leaking Ignition timing incorrect Temperature sensor fault Load sensor fault Low tyre pressures Driving style!
Fuel leakage	Damaged pipes or unions Fuel tank damaged Tank breathers blocked
Fuel smell	Fuel leak Breather incorrectly fitted Fuel cap loose Engine flooding
Incorrect emissions	Incorrect adjustments Fuel system fault Air leak into inlet Blocked fuel filter Blocked air filter Ignition system fault

Exhaust

Fault or symptom	Possible cause
Excessive noise	Leaking exhaust system or manifold joints Hole in exhaust system

Fault or symptom	Possible cause
Excessive fumes in car	Leaking exhaust system or manifold joints
Rattling noise	Incorrect fitting of exhaust system Broken exhaust mountings Engine mountings worn

Engine management

The following table is based on those available from 'Autodata' in their excellent range of books. It relates in particular to the Bosch LH-Jetronic fuel system, but it is also a good guide to many other systems. The numbers relate to the order in which the systems should be checked.

Fault or symptom	Possible cause
Engine will not start	1. Engine and battery earth connections 2. Fuel filter and fuel pump 3. Air intake system for leaks 4. Fuses/fuel pump/system relays 5. Fuel injection system wiring and connections 6. Coolant temperature sensor 7. Auxiliary air valve/idle speed control valve 8. Fuel pressure regulator and delivery rate 9. ECU and connector 10. Limp home function – if fitted
Engine difficult to start when cold	1. Engine and battery earth connections 2. Fuel injection system wiring and connections 3. Fuses/fuel pump/system relays 4. Fuel filter and fuel pump 5. Air intake system for leaks 6. Coolant temperature sensor 7. Auxiliary air valve/idle speed control valve 8. Fuel pressure regulator and delivery rate 9. ECU and connector 10. Limp home function – if fitted
Engine difficult to start when warm	1. Engine and battery earth connections 2. Fuses/fuel pump/system relays 3. Fuel filter and fuel pump 4. Air intake system for leaks 5. Coolant temperature sensor 6. Fuel injection system wiring and connections 7. Air mass meter 8. Fuel pressure regulator and delivery rate 9. Air sensor filter 10. ECU and connector 11. Knock control – if fitted
Engine starts then stops	1. Engine and battery earth connections 2. Fuel filter and fuel pump 3. Air intake system for leaks 4. Fuses/fuel pump/system relays 5. Idle speed and CO content 6. Throttle potentiometer 7. Coolant temperature sensor 8. Fuel injection system wiring and connections 9. ECU and connector 10. Limp home function – if fitted
Erratic idling speed	1. Engine and battery earth connections 2. Air intake system for leaks 3. Auxiliary air valve/idle speed control valve 4. Idle speed and CO content 5. Fuel injection system wiring and connections

Fault or symptom	Possible cause	
	6.	Coolant temperature sensor
	7.	Knock control – if fitted
	8.	Air mass meter
	9.	Fuel pressure regulator and delivery rate
	10.	ECU and connector
	11.	Limp home function – if fitted
Incorrect idle speed	1.	Air intake system for leaks
	2.	Vacuum hoses for leaks
	3.	Auxiliary air valve/idle speed control valve
	4.	Idle speed and CO content
	5.	Coolant temperature sensor
Misfire at idle speed	1.	Engine and battery earth connections
	2.	Air intake system for leaks
	3.	Fuel injection system wiring and connections
	4.	Coolant temperature sensor
	5.	Fuel pressure regulator and delivery rate
	6.	Air mass meter
	7.	Fuses/fuel pump/system relays
Misfire at constant speed	1.	Air flow sensor
Hesitation when accelerating	1.	Engine and battery earth connections
	2.	Air intake system for leaks
	3.	Fuel injection system wiring and connections
	4.	Vacuum hoses for leaks
	5.	Coolant temperature sensor
	6.	Fuel pressure regulator and delivery rate
	7.	Air mass meter
	8.	ECU and connector
	9.	Limp home function – if fitted
Hesitation at constant speed	1.	Engine and battery earth connections
	2.	Throttle linkage
	3.	Vacuum hoses for leaks
	4.	Auxiliary air valve/idle speed control valve
	5.	Fuel lines for blockage Fuel filter and fuel pump
	6.	Injector valves
	7.	ECU and connector
	8.	Limp home function – if fitted
Hesitation on over run	1.	Air intake system for leaks
	2.	Fuel injection system wiring and connections
	3.	Coolant temperature sensor
	4.	Throttle potentiometer
	5.	Fuses/fuel pump/system relays
	6.	Air sensor filter
	7.	Injector valves
	8.	Air mass meter
Knock during acceleration	1.	Knock control – if fitted
	2.	Fuel injection system wiring and connections
	3.	Air mass meter
	4.	ECU and connector
Poor engine response	1.	Engine and battery earth connections
	2.	Air intake system for leaks
	3.	Fuel injection system wiring and connections
	4.	Throttle linkage
	5.	Coolant temperature sensor
	6.	Fuel pressure regulator and delivery rate
	7.	Air mass meter
	8.	ECU and connector
	9.	Limp home function – if fitted

Fault or symptom	Possible cause
Excessive fuel consumption	1. Engine and battery earth connections
	2. Idle speed and CO content
	3. Throttle potentiometer
	4. Throttle valve/housing/sticking/initial position
	5. Fuel pressure regulator and delivery rate
	6. Coolant temperature sensor
	7. Air mass meter
	8. Limp home function – if fitted
CO level too high	1. Limp home function – if fitted
	2. ECU and connector
	3. Emission control and EGR valve – if fitted
	4. Fuel injection system wiring and connections
	5. Air intake system for leaks
	6. Coolant temperature sensor
	7. Fuel pressure regulator and delivery rate
CO level too low	1. Engine and battery earth connections
	2. Air intake system for leaks
	3. Idle speed and CO content
	4. Coolant temperature sensor
	5. Fuel injection system wiring and connections
	6. Injector valves
	7. ECU and connector
	8. Limp home function – if fitted
	9. Air mass meter
	10. Fuel pressure regulator and delivery rate
Poor performance	1. Engine and battery earth connections
	2. Air intake system for leaks
	3. Throttle valve/housing/sticking/initial position
	4. Fuel injection system wiring and connections
	5. Coolant temperature sensor
	6. Fuel pressure regulator/fuel pressure and delivery rate
	7. Air mass meter
	8. ECU and connector
	9. Limp home function – if fitted

Clutch

Fault or symptom	Possible cause
No pedal resistance	Broken cable
	Air in hydraulic system
	Hydraulic seals worn
	Release bearing or fork broken
	Diaphragm spring broken
Clutch does not disengage	As above; or
	Disc sticking in gearbox splines
	Disc sticking to flywheel
	Faulty pressure plate
Clutch slip	Incorrect adjustment
	Worn disc linings
	Contaminated linings (oil or grease)
	Faulty pressure plate
Judder when engaging	Contaminated linings (oil or grease)
	Worn disc linings
	Distorted or worn pressure plate
	Engine mountings worn, loose or broken
	Clutch disc hub splines worn

Fault or symptom	Possible cause
Noisy operation	Broken components Release bearing seized Disc cushioning springs broken
Snatching	Disc cushioning springs broken Operating mechanism sticking (requires lubrication?)

Manual gearbox

Fault or symptom	Possible cause
Noisy in a particular gear (with engine running)	Damaged gear Worn bearing
Noisy in neutral (with engine running)	Input shaft bearings worn (goes away when clutch is pushed down?) Lack of lubricating oil Clutch release bearing worn (gets worse when clutch is pushed down?)
Difficult to engage gears	Clutch problem Gear linkage worn or not adjusted correctly Worn synchromesh units Lack of lubrication
Jumps out of gear	Gear linkage worn or not adjusted correctly Worn selector forks Detent not working Weak synchromesh units
Vibration	Lack of lubrication Worn bearings Mountings loose
Oil leaks	Gaskets leaking Worn seals

Automatic transmission

Fault or symptom	Possible cause
Fluid leaks	Gaskets or seals broken or worn Dip stick tube seal Oil cooler or pipes leaking
Discoloured and/or burnt smell to fluid	Low fluid level Slipping clutches and/or brake bands in the gearbox Fluid requires changing
Gear selection fault	Incorrect selector adjustment Low fluid level Incorrect kick down cable adjustment Load sensor fault (maybe vacuum pipe, etc.)
No kickdown	Incorrect kick down cable adjustment Kick down cable broken Low fluid level
Engine will not start or starts in gear	Inhibitor switch adjustment incorrect Faulty inhibitor switch Incorrect selector adjustment
Transmission slip, no drive or poor quality shifts	Low fluid level Internal autobox faults often require the attention of a specialist

Stall test

To assist with the diagnosis of automatic transmission faults, a stall test is often used. The duration of a stall test must not be more than about 7 seconds. You should also allow about 2 minutes before repeating the test. If the precautions mentioned are not observed, the gearbox will overheat.

The function of this test is to determine the correct operation of the torque converter, and that there is no transmission clutch slip.

- run engine up to normal operating temperature by road test if possible
- check transmission fluid level and adjust if necessary
- connect a rev counter to the engine
- apply handbrake and chock the wheels
- apply foot brake, select 'D' and fully press down the throttle for about 7 seconds
- note the highest rev/min obtained (2500 to 2750 is a typically acceptable range)
- allow 2 minutes for cooling and then repeat the test in '2' and 'R'.

Final drive

Fault or symptom	Possible cause
Oil leaks	Gaskets split Drive shaft oil seals Final drive output bearings worn (drive shafts drop and cause leaks)
Noisy operation	Low oil level Incorrect pre-load adjustment Bearings worn
Whining noise	Low oil level Worn differential gears

Drive shafts

Fault or symptom	Possible cause
Vibration	Incorrect alignment of propshaft joints Worn universal or CV joints Bent shaft Mountings worn
Grease leaking	Gaiters split or clips loose
Knocking noises	Dry joints Worn CV joints (gets worse on tight corners)

Suspension

Fault or symptom	Possible cause
Excessive pitching	Defective dampers Broken or weak spring Worn or damaged anti-roll bar mountings
Wandering or instability	Broken or weak spring Worn suspension joints Defective dampers
Wheel wobble	Worn suspension joints
Pulling to one side	Worn suspension joints Accident damage to suspension alignment
Excessive tyre wear	Worn suspension joints Accident damage to suspension alignment Incorrect trim height (particularly hydrolastic systems)

Steering, wheels and tyres

Fault or symptom	Possible cause
Wandering or instability	Incorrect wheel alignment Worn steering joints Wheels out of balance Wheel nuts or bolts loose
Wheel wobble	Front or rear Wheels out of balance Damaged or distorted wheels/tyres Worn steering joints
Pulling to one side	Defective tyre Excessively worn components Incorrect wheel alignment
Excessive tyre wear	Incorrect wheel alignment Worn steering joints Wheels out of balance Incorrect inflation pressures Worn dampers Driving style!
Excessive free play	Worn track rod end or swivel joints Steering column bushes worn Steering column universal joint worn
Stiff steering	Lack of steering gear lubrication Seized track rod end joint or suspension swivel joint Incorrect wheel alignment Damage to steering components

Brakes

Fault or symptom	Possible cause
Brake fade	Incorrect linings Badly lined shoes Distorted shoes Overloaded vehicle Excessive braking
Spongy pedal	Air in system Badly lined shoes Shoes distorted or incorrectly set Faulty drums Weak master cylinder mounting
Long pedal	Discs running out, pushing pads back Distorted damping shims Misplaced dust covers Drum brakes need adjustment Fluid leak Fluid contamination Worn or swollen seals in master cylinder Blocked filler cap vent
Brakes binding	Brakes or handbrake maladjusted No clearance at master cylinder push rod Seals swollen Seized pistons Shoe springs weak or broken Servo faulty
Hard pedal – poor braking	Incorrect linings Glazed linings Linings wet, greasy or not bedded correctly Servo unit inoperative Seized calliper pistons Worn dampers causing wheel bounce

Fault or symptom	Possible cause
Brakes pulling	Seized pistons Variation in linings Unsuitable tyres or pressures Loose brakes Greasy linings Faulty drums, suspension or steering
Fall in fluid level	Worn disc pads External leak Leak in servo unit
Disc brake squeal – pad rattle	Worn retaining pins Worn discs No pad damping shims or springs
Uneven or excessive pad wear	Disc corroded or badly scored Incorrect friction material
Brake judder	Excessive disc or drum run-out Calliper mounting bolts loose Worn suspension or steering components

Hydraulic faults

Brake hose clamps will assist in diagnosing hydraulic faults and enable a fault to be located quickly. Follow these steps:

- clamp all hydraulic flexible hoses and check the pedal
- remove the clamps one at a time and check the pedal again (each time)
- the location of air in the system or the faulty part of the system will now be apparent.

Electrical

The complexity of modern electrical systems means it is only possible to give general examples in this section. However the common systems are covered.

Fault or symptom	Possible cause
Battery loses charge	Defective battery Slipping alternator drive belt Battery terminals loose or corroded Alternator internal fault (diode open circuit, brushes worn or regulator fault, etc.) Short circuit component causing battery drain even when all switches are off
Charge warning light stays on when engine is running	Slipping or broken alternator drive belt Alternator internal fault (diode open circuit, brushes worn or regulator fault, etc.) Loose or broken wiring/connections
Charge warning light does not come on at any time	Alternator internal fault (brushes worn, open circuit or regulator fault, etc.) Blown warning light bulb
Lights do not work	Bulbs blown Loose or broken wiring/connections/fuse Relay not working Corrosion in light units Switch not making contact
Instruments do not work or are inaccurate	Loose or broken wiring/connections/fuse Inoperative instrument voltage stabiliser Sender units (sensor) faulty Gauge unit fault (not very common)

Fault or symptom	Possible cause
Horn not working or poor sound quality	Loose or broken wiring/connections/fuse Corrosion in horn connections Switch not making contact High resistance contact on switch or wiring Relay not working
Wipers not working or poor operation	Loose or broken wiring/connections/fuse Corrosion in wiper connections Switch not making contact High resistance contact on switch or wiring Relay/timer not working Motor brushes or slip ring connections worn Blades and/or arm springs in poor condition
Washers not working or poor operation	Loose or broken wiring/connections/fuse Corrosion in washer motor connections Switch not making contact Pump motor poor or not working Blocked pipes or jets Incorrect fluid additive used
Electric windows not working or poor operation	Loose or broken wiring/connections/fuse Corrosion in connections Switch not making contact High resistance contact on switch or wiring Relay/timer not working Motor brushes worn
Central locking not working or poor operation	Loose or broken wiring/connections/fuse Switch not making contact High resistance contact on switch or wiring Relay/timer not working Actuators not working Remote key battery low
Heated rear window not working or poor operation	Loose or broken wiring/connections/fuse Switch not making contact High resistance contact on switch or wiring Relay/timer not working Heater element broken
Heater blower not working or poor operation	Loose or broken wiring/connections/fuse Switch not making contact Motor brushes worn Speed selection resistors open circuit
ICE system not working or poor operation	Loose or broken wiring/connections/fuse Damaged speakers Aerial damaged or not making good earth connection Interference suppression not fitted or not working Dirty cassette heads
Starter not working or poor operation	Low battery, etc. See also 'Engine' diagnostics section above

LEARNING TASKS

➡ Look back at the key words. Explain each one to a friend, and/or write out a short description to keep as evidence.

➡ Examine a real system and note how the contents of this chapter could be used to help with fault finding.

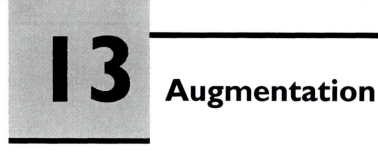

13 Augmentation

1 Introduction

Start here! So what does this strange word *'augment'* mean?

> "**aug ment** (og ment; *for n.* og ment) **vt.** [[ME augmenten < OFr augmenter < LL augmentare < augmentum, an increase < L augere, to increase: see WAX²]] I to make greater, as in size, quantity, or strength; enlarge 2 to add an augment to **vi.** to become greater; increase **n.** I [Obs.] an increase 2 Gram. a prefixed vowel or a lengthening or diphthongization of the initial vowel to show past time in Greek and Sanskrit verbs **SYN.** INCREASE aug ment able *adj.* **aug ment er** *n.*".
>
> (Excerpted from *Compton's Interactive Encyclopedia.* Copyright © 1994, 1995 Compton's NewMedia, Inc.)

That's made it a lot clearer!

Seriously though, in our situation augmentation can be described simply as *'carrying out work to improve or enhance a vehicle'*. This work can range from fitting mud flaps to race-tuning an engine. Some of the main areas you may come across are examined later in this chapter. Here are some advantages and disadvantages of augmentation:

KEY WORDS

- All words in the table
- Instructions
- Customers
- Accessories

Advantages	Disadvantages
Improve the look of the vehicle	Expensive
Increase performance	Could make the car more likely to be stolen
Better lighting output	Components fitted may not transfer to any other subsequent vehicle
Improved sound system	Engine performance may exceed the handling abilities
Enhance the resale value	
Easier to carry loads	
Better handling	

Figure 13.1 shows some accessories available from Peugeot for vehicle augmentation.

Terminology	
Retrofit	An accessory or component designed to be fitted in the after-sales market
Accessories	Any extra component fitted to a vehicle, for example, spot lights
ICE	In-car entertainment
Factory fitted	Accessories or components fitted as the vehicle was being assembled on the line
Recommended accessories	Some manufacturers only recommend their own or specific items to be fitted. In some cases, if this recommendation is not followed the warranty may be made invalid
BS	British Standard
ISO	International Standards Organisation
Safety	The most important issue when fitting some components to augment a vehicle, for example, child seats

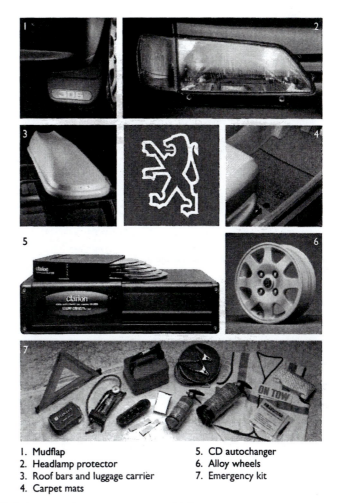

1. Mudflap
2. Headlamp protector
3. Roof bars and luggage carrier
4. Carpet mats
5. CD autochanger
6. Alloy wheels
7. Emergency kit

Figure 13.1 Accessories to augment a vehicle

Advising customers

Advising customers is one of the most important jobs we do. We must also be able to interpret their instructions. Equally, though, it is not always appropriate to just blindly do whatever the customer tells you to do. For example, let's assume you are told by the customer that s/he wants a certain size of wheels and tyres fitted to his/her car. If you carry this job out without thinking, a serious accident could occur, if say, the tyres rubbed on the wheel arch as the car hit a bump, and went out of control. When brought to court, it is quite likely that the customer will blame you for not advising them correctly. This might sound a bit extreme but it does happen. Remember you are now becoming thought of as an expert, and you will be expected to act like one.

It is not possible for me to cover all eventualities following all the different types of augmentation you may carry out. As usual, common sense and a good knowledge of the vehicle is the key to success.

Following instructions

'If at first you don't succeed, look at the instructions!' We've all done it at one time. Of course it is wrong, you should refer to the instructions at all times when carrying out any repairs, not just augmentations. It is also best to read all the instructions before you start. As an exercise, carry out the following:

1. Turn back to the start of this chapter and read the definition of augmentation.
2. Close the book.
3. Write out all you can remember without looking.
4. Make a sketch to show a car with parts fitted to augment its looks.
5. Do not carry out parts one to four.

All right, it was a cheap trick but it illustrates a point! As a further example of how you must follow *all* instructions, read those for fitting a tow bar later in this chapter. What would happen if you did not fit the large flat washers? They are only small parts, but very important to spread the load over a wider area.

Read the instructions

LEARNING TASKS

➡ Look back at the key words. Explain each one to a friend, and/or write out a short description to keep as evidence.

➡ Examine a real vehicle and make a list of possible augmentations — you may assume cost is not a problem!

2 Types of augmentation

In-car entertainment (ICE)

It would now be almost unthinkable not to have a radio cassette player in our vehicles. It does not seem too long ago, however, that these were an optional extra! We now have in-car entertainment systems fitted to standard production cars which are of very good hi-fi quality. Facilities such as compact disc players and multiple compact disc changers, together with automatic station search and re-tune, are becoming ever more popular.

KEY WORDS

■ All words in the tables

Power supplies

Figure 13.2 shows the connections required to operate a quality ICE system. Fuses of the correct ratings should be used (refer to manufacturer's specifications). An electric aerial is included as well as the connection to a multi-compact disc unit via a data bus

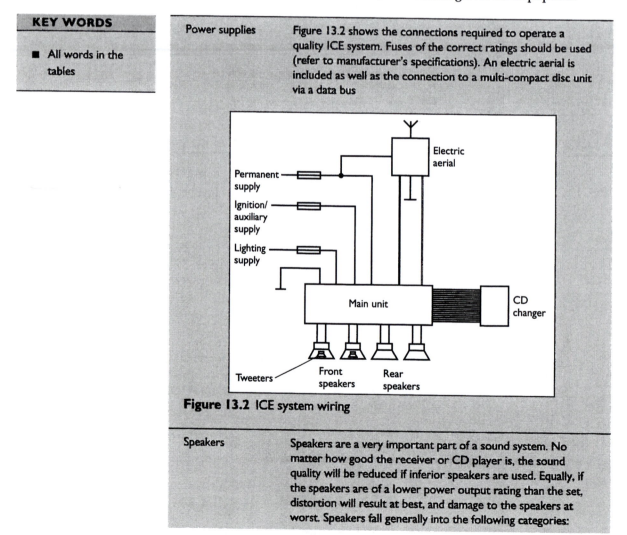

Figure 13.2 ICE system wiring

Speakers

Speakers are a very important part of a sound system. No matter how good the receiver or CD player is, the sound quality will be reduced if inferior speakers are used. Equally, if the speakers are of a lower power output rating than the set, distortion will result at best, and damage to the speakers at worst. Speakers fall generally into the following categories:

Tweeters – high frequency reproduction

Mid range – middle range frequency reproduction (treble)

Woofers – low frequency reproduction (bass)

Sub-woofers – very low frequency reproduction

Figure 13.3 shows the construction of a speaker and Figure 13.4 shows a sub-woofer produced by Pioneer®

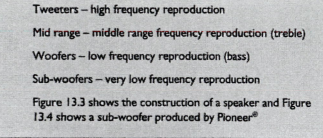

PPTA high-range diaphragm
Compact midrange/tweeter design
Metal-coated injection-moulded polypropylene (IMPP) cone
Resonant spacer

Triangular butyl rubber edge
Full depth basket
Large diameter conex damper
Long voice coil design

Figure 13.3 Speaker construction

Figure 13.4 Pioneer sub-woofer

Wiring connections	Connections should always be of high quality; in other words secure, clean and with a low electrical resistance. Proper crimp terminals fixed with the proper tool, or soldered joints are the best methods
Security	High quality systems can be very expensive. Some can be linked to the vehicle alarm system. Many now have removable front covers, without which the system is worthless
Secure fitting	Ensure the mountings are secure because vibration can affect sound quality or even damage the system
Specialist systems	Figure 13.5 shows an example of some top range system components

Multiple changer system

Head unit
P-series

CD-P33
adaptor

Up to 3 P-series Multi-CD players
(CDX-P5000, CDX-P2000,
CDX-P1220S or CDX-P620S)

Figure 13.5 ICE system

Interference suppression	Most modern vehicles are well suppressed as standard. However, vehicle electrical systems produce a lot of interference noise. This is discussed in a later section
Instructions to the customer	Ensure you pass on any manuals or instruction sheets to the customer. In many cases, you will be expected to help them understand how the system works. Remember that misunderstandings can cause customer dissatisfaction which, with a little thought, can be easily prevented

Most good ICE systems now include at least six speakers, two larger speakers in the rear parcel shelf to produce good low frequency reproduction, two front door speakers for mid range and two front door tweeters for high frequency notes. Controls on the set will include volume, treble, bass, balance and fade. Cassette tape options will include 'Dolby' filters to reduce hiss, and other tape selections such as chrome or metal. A digital display of course, will provide a visual output of operating condition. This is also linked into the vehicle lighting to prevent glare at night. Track selection and programming for one or several compact discs is possible.

An interesting development is RDS, which stands for 'radio data system'. This is an extra inaudible digital signal, which is sent with FM broadcasts in a similar way to how teletext is sent with TV signals. RDS provides all the information a receiver needs to act intelligently. RDS allows the station name to be displayed in place of the frequency, automatic tuning to the best available signal for the chosen radio station, and traffic information broadcasts to be identified. A setting can even be made so that whatever you are listening to at the time can be interrupted.

ICE system augmentations will probably involve replacing existing systems. There are, therefore, a number of important points to be observed as noted in the previous table.

Mobile communications

If the success of the cellular industry is any indication of how much use we can make of the telephone, the future promises an even greater expansion.

The desire to keep in touch with each other is so great that an increasing number of people now have up to five phone numbers: home, office, pager, fax and cellular. However, within the foreseeable future, high-tech digital radio technology and sophisticated telecommunications systems will enable all communications to be processed through a single number. With this 'personal numbering' a person carrying a pocketsize telephone will need only one telephone number. Instead of people calling places, people will call people; we will not be tied to any particular place. Personal numbering will make business people more productive. They will be able to reach and be reached by colleagues and clients, anywhere and anytime, indoors or outdoors. When travelling from home to office or from one meeting to the next, it will be possible to communicate with anyone, whenever the need arises. Yet, if a user wishes to be temporarily inaccessible, it will be a simple matter of turning off the telephone.

CB radios and short range, two-way systems such as those used by taxi firms and service industries will still have a place for the time being. Even these may decline as the cellular network becomes cheaper and more convenient to use.

Nevertheless, where does this leave communication systems relating to the vehicle? Well, it is the opinion of many that 'in vehicle' communication equipment for normal business and personal use will be by the simple, pocket-sized mobile phone. This means that there is little further market for the car telephone.

One important area of work remaining is the 'hands free' conversion. 'Hands free' car telephones allow both hands to be kept free to control the car. This is an augmentation *and* a safety issue.

There are a number of important points to be observed when fitting a 'hands free' conversion:

Power supplies	Fuses of the correct ratings should be used (refer to manufacturer's specifications)
Speakers	Often the existing ICE system speakers can be used but if not, general purpose speakers are adequate because high quality reproduction is not necessary
Wiring connections	Connections should always be of high quality; in other words secure, clean and with a low electrical resistance. Proper crimp terminals fixed with the proper tool, or soldered joints are the best methods

Security	The mobile phone is often fitted into a holder on the vehicle's central console or fascia. The customer should be reminded not to leave the phone on view when s/he leaves the vehicle
Secure fitting	Ensure all mountings are secure because vibration can damage the system or prevent correct operation
Instructions to the customer	Ensure you pass on any manuals or instruction sheets to the customer. In many cases, you will be expected to help them understand how the system works. Remember that misunderstandings can cause customer dissatisfaction which, with a little thought, can be easily prevented

Interference suppression

The process of interference suppression is simply to reduce the amount of unwanted noise produced from the speakers of the ICE or communications system. Interference can get into the sound or communications system of a vehicle in two ways:

■ line borne – through the wires
■ air borne – through the air via the aerial.

The sources of interference in the motor vehicle can be summarised quite simply as any circuit which is switched or interrupted suddenly. This includes the action of a switch and the operation of a motor. Interference is produced from four main areas of the vehicle:

■ ignition system
■ charging system
■ motors and switches
■ static discharges.

The ignition system of a vehicle is the largest source of interference, in particular the high-tension side. Voltages up to 30 000 V are now common. The charging system produces noise because of the sparking at the alternator brushes. Any motor, switch or relay, is likely to produce some interference. The most common sources are the wiper motor and heater motor. The starter is not considered due to its short usage time.

Build up of static electricity is due to friction between the vehicle and the air, and the tyres and the road. If the static on the bonnet builds up more than the wing, then a spark can be discharged. This is easily prevented by using bonding straps to ensure all panels stay connected. There are four main techniques for suppressing radio interference:

■ Resistance is used in the ignition HT circuit, up to a maximum of about 20 kΩ per lead. This has the effect of limiting peak current, which in turn limits electromagnetic radiation. The spark quality is not affected.
■ As mentioned earlier, bonding is simply ensuring all parts of the vehicle are electrically connected together. This will prevent sparking due to the build up of static.
■ Screening is generally used only for specialist applications such as emergency services and the military. It involves completely enclosing the ignition system and other major sources of noise in a conductive screen connected the vehicle earth. This prevents interference waves escaping. It is a very effective, but expensive technique.
■ Capacitors and inductors are used to act as filters. By choosing suitable values of capacitors in parallel and/or inductors in series, it is possible to filter out unwanted noise signals. Capacitors are often fitted on the alternator main output and the power supply side of the ignition coil.

Vehicle security

Alarmed? You should be! More than several hundred thousand cars are stolen in the UK every year. Alarm systems in general are discussed in chapter 11. However, other techniques can be used to improve vehicle security.

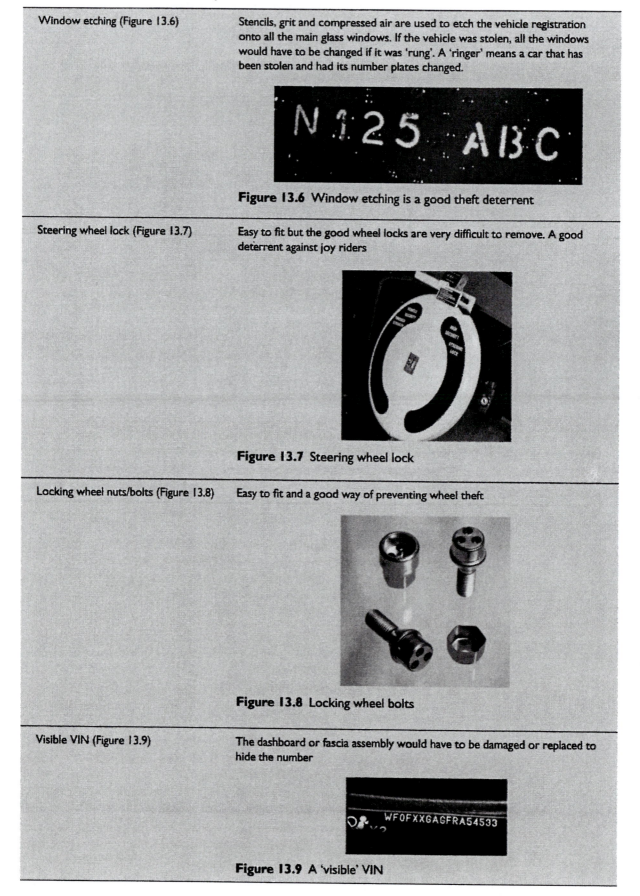

Window etching (Figure 13.6)

Stencils, grit and compressed air are used to etch the vehicle registration onto all the main glass windows. If the vehicle was stolen, all the windows would have to be changed if it was 'rung'. A 'ringer' means a car that has been stolen and had its number plates changed.

Figure 13.6 Window etching is a good theft deterrent

Steering wheel lock (Figure 13.7)

Easy to fit but the good wheel locks are very difficult to remove. A good deterrent against joy riders

Figure 13.7 Steering wheel lock

Locking wheel nuts/bolts (Figure 13.8)

Easy to fit and a good way of preventing wheel theft

Figure 13.8 Locking wheel bolts

Visible VIN (Figure 13.9)

The dashboard or fascia assembly would have to be damaged or replaced to hide the number

Figure 13.9 A 'visible' VIN

Body fittings The range of fittings is quite large, from the simple application of a badge or transfer to fitting roof racks or full body spoiler kits. Figures 13.10, 13.11 and 13.12 show some examples. It is not possible to cover all the situations you will come across, but the following list is good general advice:

- Take extra care not to damage paintwork. Use covers and tape if necessary
- Follow instructions
- Use only recommended accessories
- Ensure the parts are fitted securely
- Make sure instruction leaflets are passed on to the customer
- Do not break rules or regulations.

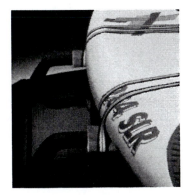

Figure 13.10 Roof rack for a surf-board

Figure 13.11 Bike roof rack fittings are popular

Figure 13.12 Rear spoilers can look attractive

Auxiliary lamps The most common auxiliary lamps you may be required to fit are listed in the following table:

High level brake lights	The use of LED's as high level brake lights is now quite popular. This is partly due to their shock resistance, which allows them to be mounted on the boot lid, and partly due to their faster illumination time. LED's reach visible brightness faster than conventional bulbs. The reduced reaction time of a following driver can equate to a car's length at motorway speeds. The normal procedure is to fit one light unit centrally in the rear window
Rear fog lights	One or two may be fitted, but if only one, it must be on the offside or centre line of the vehicle. They must be between 250 mm and 1000 mm above the ground and over 100 mm from any brake light. The wattage is normally 21 W, and they must only operate when either the side lights, headlights or front fog lights are in use
Front spot and fog lights	If front spot lights are fitted (auxiliary driving lights, Figure 13.13), they must be between 500 mm and 1200 mm above the ground and more than 400 mm from the side of the vehicle. If the lights are non-dipping then they must only operate when the headlights are on main beam. Front fog lamps are fitted below 500 mm from the ground, and may only be used in fog or falling snow. Spot lamps are designed to produce a long beam of light to illuminate the road in the distance. Fog lights are designed to produce a sharp cut-off line such as to illuminate the road just in front of the vehicle but without reflecting back or causing glare. The wattage varies from 55 W up to 100 W or more

Figure 13.13 Auxiliary driving lamps

Extra points to note about fitting auxiliary lights:

- use the correct size cable – each strand will carry about 0.5 A
- use a relay for spot and fog lights – see Figure 13.14
- ensure the lights operate to comply with regulations – for example, spot lights only on with main beam (Figure 13.14)
- before fitting extra wiring, check to see if connections are already available.

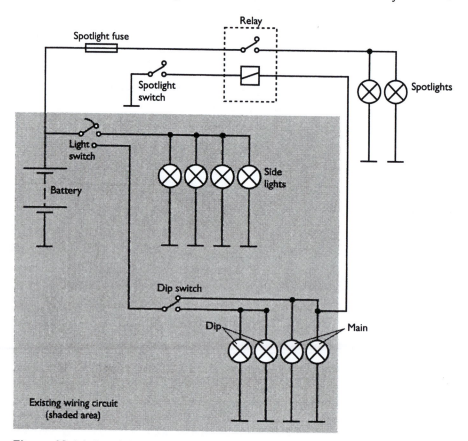

Figure 13.14 Spot light wiring using a relay

Wheels and tyres

Fitting new wheels and tyres is often a way of making a vehicle look good (Figure 13.15). There are now so many good alloy wheels available that we are spoilt for choice. The following is a useful checklist, to be referred to when fitting new wheels and/or tyres that are other than standard:

- follow manufacturer's recommendations
- torque wheel nuts to the correct value
- ensure the tyres do not foul the vehicle body under any circumstances
- wheels or tyre should not protrude beyond the wheel arch
- tyres with a different diameter will affect the speedometer reading
- is the spare tyre still useable?
- will the vehicle handling be affected?

Figure 13.15 Alloy wheels and lockable wheel nuts

Performance modifications   For the purpose of this section, we will look briefly at just three areas relating to performance modification:

- engine
- suspension
- brakes.

The subject of performance modifications is very interesting but also far-reaching. It would require a complete book just to look at the issues concerned with performance modifications to just one type of vehicle. For this reason, the subject of this section will just look at some of the *possible* advantages and disadvantages (where appropriate) of selected modifications. The following three tables look at engine, suspension and braking modifications (cost has been ignored but is clearly an issue for you or your customer to consider):

Engine modifications	Advantages	Disadvantages
Four branch exhaust manifold	Improved gas flow out of the combustion chamber	Could be difficult to fit because of space limitations
High lift camshaft	Improved gas flow due to more valve lift during the open time. The mixture can enter and the exhaust can leave more efficiently	High wear on other valve components
Skimmed head	Higher compression ratio may improve performance	Increased stress on the engine and problems may be encountered with valve timing due to the fitting of the cam belt/chain (OHC)
Gas flowed head	'Porting and polishing' is the process of matching the manifolds to the head and ensuring the walls are smooth. This can improve gas flow and, hence, performance	In some cases, it is necessary to have rough surfaces to increase turbulence!
Hot chipping	Changing the timing and fuelling characteristics by changing the memory configuration of the engine management ECU will improve operation of the engine performance-wise	The increased performance is at the expense of fuel consumption, exhaust emissions and engine life
Engine balancing	Smoother operation of the engine, particularly at higher speeds	None
Up rated components	Improved strength and quality of engine components will make the system last much longer and run more smoothly	None
All of the above	Significant improvement in engine power and torque Very impressive for race circuit use	High fuel consumption. In addition, the rest of the vehicle may need modifications Not really appropriate to road cars

Suspension modifications	Advantages	Disadvantages
Stiffer or progressive springs	Better high speed and rough condition handling	Harder ride if springs are just stiffened but progressive springs may be a good compromise
Lowering the ride height	Looks good	May make the handling poor or even very dangerous. Steering will be affected
Up rated or adjustable dampers	Excellent for appropriate conditions such as off-road use	Harsher ride under normal conditions
Changes to camber and swivel axis inclination	Possible improved operation and handling	Tyre wear
All of the above	Significant handling improvements, particularly for circuit conditions	Resulting ride may not be ideal for general use

Braking modifications	Advantages	Disadvantages
Carbon-kevlar pads	Improved performance – movement energy is converted to heat more rapidly	Standard discs can warp from the extra heat produced
Ventilated discs	Better cooling of the discs giving improving braking	None
Larger/more powerful servo	Higher braking pressures and, hence, performance	Extra heat generated and serious risk of hydraulic component failure
Up-rated components	Safer operation	None
All of the above	Significant improvement in performance of the brakes	May react too harshly for normal road use

Much of what has been discussed briefly here is dealt with further in the technology chapters of the books. However, bear some important points in mind:

- a vehicle operates as a system, modifying one part may adversely affect another
- a series of modifications may be necessary before any benefit is noticed. For example, fitting a high lift cam will make very little difference unless the fuelling and exhaust systems are also modified
- a vehicle with a high performance engine should also have high performance brakes and suspension!
- a noisy modified exhaust can actually reduce performance due to incorrect back pressures
- making any component work towards its tolerances will reduce its life significantly
- fitting 'go faster stripes' on the side of a car is a lot cheaper than really making it go faster!

Tow bars and towing The following is a list of instructions for fitting a Witter® tow bar to a light vehicle:

Materials

A	1	Tubular cross bar with international facing to ISO 3853 (BS AU114)
C1	2	Side Mounting Arms (1 off each hand)
D	4	M10 × 110 mm bolts, nuts, lock washers and flat washers (30 mm O.D.)
D1	4	Bushes (17 mm O.D. – 79 mm long)
E	4	M12 × 35 mm bolts, nuts, lock washers
CB	2	M16 × 50 mm coupling bolts, nuts, lock washers

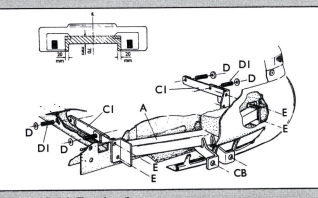

Figure 13.16 Tow bar fitting

Fitting

1. Remove the bumper (2 mud flap screws, 2 tip screws behind the inner-shield at the top, 2 nuts inside the inner wings, 3 Torx head screws along top edge and one clip underneath bumber, 4 nuts on rear panel). Cut the section out of the foam bumper insert as shown.
2. Slip the rubber exhaust hanger off its support and lower the exhaust heat shield to gain access to the chassis rails. Insert bushes (D1) on bolts (D) with large flat washers under bolt heads into the frame through the adjacent 'keyholes' and loosely attach side arms (C1).
3. Drill out the rear holes in the frame for the M12 bolts (E) and fit the main cross bar (A). Tighten all bolts.
4. Replace the bumper and its foam insert before refitting the heat shield and exhaust.

Further information and issues you should be aware of are listed below.

- Some models have the number plate below the bumper. Ensure that the number plate is not obscured. Raise it as high as possible on its mounting and/or remove the ball when not towing.
- Refer to the vehicle specification for the trailer weight and nose load limits, which **must be observed**. As a guide, 50 kg imposed on the tow ball by the trailer or caravan is a good average.
- Any damage due to accident or corrosion should be repaired before fitting the towing bracket.
- Torque settings, if given by the tow bar manufacturer, should be followed. If not available, the following is a good guide for high tensile bolts: M8 = 29 Nm, M10 = 58 Nm, M12 = 102 Nm and M16 = 252 Nm.

Information you might provide for the customer:

- the security of the tow bar should be checked periodically
- the trailer or caravan should be correctly loaded
- using a stabiliser will help to reduce the effect of cross winds and overtaking vortices from large vehicles. However, a stabiliser should not be used to make up for a badly loaded trailer or caravan.

Electrical socket wiring is often required in conjunction with fitting a tow bar. Many manufacturers now provide a plug and socket connection to make tow bar wiring easy. If this is not the case then the connections in Figure 13.17 should be followed.

Terminal	Colour	12	12S
1	Yellow	LH indicator	Reversing light
2	Blue	Rear fog lamp	Battery charging
3	White	Common return	Common return
4	Green	RH indicator	Power supply
5	Brown	RH side/tail	Sensing device
6	Red	Stop lamp	Refrigerator
7	Black	LH side/tail	Spare

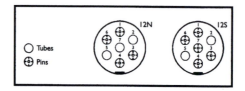

Figure 13.17 Tow bar wiring

Information sources   In an earlier section, it was suggested that you should always read the instructions! Further to this, you will gain valuable information from the following sources:

- sales literature
- technical data
- fitting procedures
- legislation.

Each of these sources will be useful for finding a different type of information. The next table lists a few suggestions:

Sales literature	This is used mostly as a pre-sales aid to help the customer decide what they would like. It is also useful to you as it will probably contain a guide as to which vehicles the 'xyz' device will fit, or which particular 'xyz' device you would need to get for the particular vehicle
Fitting procedures	These are the instructions as discussed earlier! Following fitting procedures will often save you time in the long run
Technical data	A good example of this would be the ratings of ICE system speakers. By referring to technical data (supplied by the manufacturer), you can check which speakers will be suitable to handle the system output
Legislation	Remember at all times that augmentations must always comply with current legislation. For example, engine modifications that cause the vehicle to fail the MOT exhaust emission limits will not be satisfactory. Another example would be incorrectly adjusted spot or fog lights

LEARNING TASKS

➡ Look back at the key words. Explain each one to a friend, and/or write out a short description to keep as evidence.

➡ Make sketches to show how to wire up spot lights with a relay.

➡ Write a short notes about advantages and disadvantages of the augmentations suggested in each section heading.

14 Valeting

1 Introduction

Start here! What is valeting? Cleaning a vehicle doesn't seem to be a good enough description. We all wash our cars regularly; I do mine every six months, whether it's dirty or not! Valeting is *professional* cleaning of a vehicle. It involves using specialist cleaning products and equipment.

When valeting a light vehicle it is worth considering the reasons for such an operation. Why should a vehicle be presented to such a high standard? I suggest that there are a number of reasons:

- it enhances the value of the vehicle
- a customer's first impressions are very important
- it is only what the customer should expect
- it simply looks good!

A vehicle will be valeted for a number of reasons:

- a request from a customer
- preparation for sale
- part of a vehicle service as an incentive to the customer.

There is no magic formula for the protection and maintenance of a vehicle's appearance. However, there are certain basic steps and a sequence in executing them that will result in a top quality finish. The correct products must always be used.

Figure 14.1 A well presented vehicle even looks good in the rain!

Regardless of the vehicle surface that you are working with, it must be clean before you proceed to condition, protect or polish it. This applies to every surface of your vehicle, outside and inside. The properties and application of exterior and interior cleaning materials will be covered in detail in the following sections.

Terminology

TFR	Traffic film remover is a strong detergent for exterior cleaning
Polish	Polymer or wax-based, this is used to give a deep shine on exterior paint. It also protects the paint from sunlight and water
Dry clean fluid	Some interior cleaning products work by using a fluid that brings the dirt to the surface as it evaporates, where it can be wiped or vacuumed off
Pressure washer	A machine with a nozzle through which water is pumped under high pressure. It is ideal for cleaning the engine bay and vehicle wheels
Steam cleaner	Similar to the pressure washer except that a heating element is used to produce very hot water. This is the best tool for engine bay cleaning
Elbow grease	The real secret to producing a good finish when valeting a vehicle! This is particularly so when polishing
Wet vacuum	A 'wetvac' is ideal for interior cleaning of seats and carpets. Water with appropriate detergent is sprayed into the upholstery and then vacuumed out again

Safety and PPE   The products for vehicle valeting are chosen to do a job quickly and to a good standard, but this must never be at the expense of safety. Let's start with the obvious safety recommendations, and remember – safety precautions are for your benefit.

- only use a product for its intended purpose
- store cleaning materials safely and away from children
- use appropriate PPE
- always follow manufacturer's instructions.

Detailed data sheets are available from the manufacturers on request. This is a requirement of the COSHH regulations.

> Because of the job they have to do, valeting products are made from strong chemicals. Take extreme care.

As with most operations, personal protective equipment for valeting is common sense:

PPE	Where used
Overalls	All the time! A pair of waterproof overalls is the ideal. Remember, though, that a clean pair may be a good idea when cleaning the vehicle interior
Rubber boots	These may be a good idea when washing the vehicle but particularly so when using steam cleaners or pressure washers
Gloves	By necessity, cleaning materials consist of strong chemicals. Barrier cream should be worn at all times but gloves must be used for further protection, particularly against solvents and strong detergents
Face mask	When there is any risk of splashing chemicals in your face, a mask should be worn. A mask will protect all of your face and not just your eyes, as goggles would. Pressure washing and steam cleaning make a face mask essential, as not only cleaning products but dirt and stones may be thrown back at you

Figure 14.2 Hazardous product label

Figure 14.2 shows a hazardous product label and Figure 14.3 shows a manufacturer's data sheet (Autoglym).

Further consideration is necessary when using steam cleaners. The potential for accidents is significant if safety procedures are not followed. Be sensible with this machine and damage or accidents will not occur.

The steam cleaner has the following safety risks:

- very hot water – risk of burns or scalding
- high-pressure water – damage to the vehicle or eyes and ears, for example
- hot machinery – the heating coils get very hot and will burn
- naked flames – burning paraffin is often used to heat the water
- strong detergents – skin and eye damage
- high voltage electrical supply – serious risk due to the vicinity of water
- metal lance – gets hot but can also damage the vehicle paintwork.

LEARNING TASKS

➡ Look back at the key words. Explain each one to a friend, and/or write out a short description to keep as evidence.

➡ Make a list of all the safety 'risks' you can think of relating to valeting. Suggest how to minimise each of them.

2 Exterior cleaning

Cleaning the wheels
The first step in washing a vehicle will often be to clean the wheels, tyres and wheel arches. If brake dust is allowed to remain on the wheel surface, it can permanently bond with and etch the surface.

AUTOGLYM

SAFETY DATA SHEET	ISSUE No.	DATE
	6	22/02/96

AUTOGLYM, LETCHWORTH, ENGLAND, TELEF: INT+44 (0)1462 677766. FAX: INT+44 (0)1462 677712

SECTION 1 : IDENTIFICATION OF SUBSTANCE / PREPARATION

PRODUCT NAME	No. 12 RADIANT WAX POLISH	Ref No.	12STD/ODK1/27493
PRODUCT TYPE	Water-in-oil emulsion	Packaging	5 litre metal can
PRODUCT USE	Automotive polish		
MANUFACTURER	Autoglym, Letchworth, England. SG6 1LU	Emergency Tel No	44 (0)1462 677766
IMPORTER		Emergency Tel No	

SECTION 2 : COMPOSITION / INFORMATION ON INGREDIENTS

HAZARDOUS INGREDIENTS	CAS NUMBER	% By WEIGHT	CLASSIFICATION SYMBOL	CLASSIFICATION RISK PHRASES	OCCUPATIONAL EXPOSURE STANDARDS LTEL ppm	OCCUPATIONAL EXPOSURE STANDARDS LTEL mg/m3	OCCUPATIONAL EXPOSURE STANDARDS STEL ppm	OCCUPATIONAL EXPOSURE STANDARDS STEL mg/m3	MAXIMUM EXPOSURE LIMITS LTEL ppm	MAXIMUM EXPOSURE LIMITS LTEL mg/m3	MAXIMUM EXPOSURE LIMITS STEL ppm	MAXIMUM EXPOSURE LIMITS STEL mg/m3	NOTES
Aliphatic Hydrocarbon Solvent (<1% aromatics)	64742-48-9	20 - 50		10	150	1000	375	2160					
Anhydrous Aluminium Silicate	66402-68-4	5 - 20				10 5							TD RD
Aliphatic Hydrocarbon Solvent (<2% aromatics)	64742-47-8	1 - 5				5		10					AOM

EXAMPLE ONLY

HSE LTEL : Long-term exposure limit (8 hour Time Weighted Average); RD : Respirable Dust ; TD : Total Dust ; AOM : As Oil Mist.
HSE STEL : Short-term exposure limit (15 minute reference period) HSE MEL : Maximum exposure limit Sk : Can be absorbed through skin

SECTION 3 : HAZARDS IDENTIFICATION

INHALATION	Irritating to eyes and respiratory tract. Anaesthetic in high concentrations and may cause drowsiness, dizziness and affect the central nervous system.
SKIN CONTACT	Irritating. Frequent or prolonged contact may cause dermatitis.
EYE CONTACT	Vapour and liquid irritating to eyes
INGESTION	Due to the presence of solvent in product, small amounts aspirated into lungs during ingestion or vomiting may cause bronchopneumonia or pulmonary oedema.

SECTION 4 : FIRST AID MEASURES

INHALATION	Remove to fresh air. Get medical help.
SKIN CONTACT	Wash with soap and water. Remove contaminated clothing.
EYE CONTACT	Flush thoroughly with water. Seek medical attention.
INGESTION	Do not induce vomiting. Seek medical help immediately.

SECTION 5 : FIRE FIGHTING MEASURES

Extinguishing Media

☐ Water Fog ■ Foam ■ CO2 ■ Dry Chemical ☐ Other

Special Fire Fighting Procedures

Unusual Fire Fighting Hazards

Issue No.6 Page 1 of 2

Figure 14.3 Safety data sheet

The solution for cleaning the wheels has a higher acid level than car shampoo. This is necessary to dissolve the brake dust that accumulates on the wheel surface. Follow the directions provided by the manufacturer. The wheel cleaner will often be sprayed on and worked with a brush. Use a shampoo solution to remove the wheel cleaner. Make sure you do not allow the wheel cleaning solution to dry on the wheel and tyre surface. Many of these cleaning solutions will etch the wheel if allowed to dry without rinsing. Thoroughly spray with water to rinse off all the cleaner to complete the job.

Washing the paint

Washing the vehicle by hand with a bodywork shampoo is the best method. Do not use household detergents because they are made to remove grease. They will also remove the polish on the paint surface and accelerate the oxidation process. Use of most commercial machine car wash systems will result in some level of surface scratching. A pressure washer is an ideal tool to assist with washing.

To wash the vehicle:

- make sure that the paint surface is cool to the touch
- wash from the top down
- use either a natural fibre mitt or a shampoo sponge
- follow the instructions regarding the amount of shampoo to use
- wash the vehicle in small sections with frequent rinsing to prevent the water and the contaminants that you are removing from drying on the surface of the paint
- rinse your mitt or sponge frequently as you progress down the sides of the vehicle, since there will be more dirt and contaminants closer to the ground
- finish with a complete rinse of the entire vehicle.

There are three steps in the drying process to get the best results:

- remove the largest volume of the water from the entire vehicle with a chamois
- blow out all the channels where water can accumulate
- towel dry all surfaces, including the windows and wheels.

Open all the doors, the engine compartment and the boot to eliminate any remaining run off and water tracks. Wind the windows down a few centimetres to clean the dirt from the seal area at the top and sides of the glass.

When the drying is complete, you are ready to inspect the exterior surface. Depending on the condition of the paint, there are several different products that may be needed to produce an ideal finish. Examples of these include:

- wax and grease strippers
- auto solvents and cleaners (mineral spirits)
- tar and road oil removers
- 'clay' – to remove paint overspray and other imbedded contaminants
- abrasives and glazes – multiple grades.

After you have finished shampooing the vehicle, you are ready to take the next step. When you examine the paint, you may find that there is the need for additional cleaning. There are paint cleaning products that use solvents of various types and strengths. They remove a wide range of surface contaminants.

Wax and grease strippers

These products are used to remove old wax from the surface. This is to make sure the final polish or wax product will adhere to the paint surface. These products will normally not remove oxidation, swirl marks, road tar, water spots or tree sap. If the paint is in good condition, these products will do an excellent job of removing old wax residue. The method of applying these products is to dampen a cotton cloth with the product and clean a single panel or section of paint with back and forth motions. The product should not be allowed to dry on the paint surface and should be removed with a second cotton cloth within a minute or two.

Solvents and mineral spirits

These products are used for removing light contamination such as tar, water spotting and tree sap. Use the same technique as for the wax and grease stripper above.

Tar and road oil removers – including Methyl Ethyl Ketone

These products are harsh on the paint surface and care should be taken. They should not remain on the surface for more than a few seconds. These products may soften the paint and cause discoloration. The application technique is the same as above but with the quantity of the product reduced to the minimum possible.

Use appropriate PPE.

Abrasives, compounds and glaze products

Caution! These products remove paint from the vehicle's surface. There is no way to put the paint back on when you go too far. The range of abrasive grit can be thought of as roughly comparable to wet and dry rubbing down paper. There are compound products that will be equivalent to the highest level of wet and dry paper of about 2000 grit.

The types of paint damage that will require this product include heavy oxidation, serious water spotting, paint overspray that has bonded and several types of air pollution that contain acid components. In other words, when the paint surface itself has been damaged. The normal procedure is to cover a small area with a back and forth motion. The product should not be allowed to dry and should be removed when it becomes cloudy or tacky. Take care!

Applying and buffing polymers and waxes

There are two main products for protecting automotive paint finishes:

- polymers – synthetic, man made substances
- waxes – naturally occurring substances found in trees (and bee hives!).

There is a great deal of discussion regarding the merits of each product. There are individuals who strongly support each one. The synthetic polymers are more similar to the chemical composition of the paint itself than the 'natural' products (and what's natural about the paint?). Consider the following if you are to choose:

- cost per application
- depth of gloss
- ease of use
- durability.

KEY WORDS

- Wax
- Polymer
- Strippers
- Solvents
- Abrasives
- Compounds

The cost of most polymers will be higher than their counterpart waxes, but polymers will last up to four times longer. This durability factor is a significant consideration. There is an edge to the polymers when comparing ease of use. They will go on easier and come off faster. An additional benefit is the clarity and refractive consistency that the polymers produce on the finish surface. When comparing depth of gloss, however, high quality waxes have an edge.

All products should be applied in a front to back motion with a cotton cloth. The drying time can range from minutes to days, to achieve maximum bonding for some polymers. I suggest a product that dries in a few minutes will be most appropriate for professional valeting.

All polymer and wax products should ideally be hand finished. However, use of an orbital buffer with a high quality pad can reduce the effort of removal of the finish product. In all cases, the final step in removal should be by hand.

Glass cleaning   Products which contain ammonia do a good job of removing the dirt and film that is found on vehicle glass. One of the main problems with cleaning the inside surfaces is the build up of contaminants. These come from the decomposition of the vinyl and plastic components. This decomposition and transfer to the surface of the glass is caused by the ultraviolet action of the sun.

It is recommended that two cotton cloths are used to clean and then dry the surface. Spray the cleaner on one cloth to prevent the cleaner from getting on the surface of the dashboard as you wipe the glass. Use the second towel to dry the surface. Clean the outside first as this makes it easier to see where there are any streaks when you do the interior. Lower the door windows to get both sides of the glass, which is recessed into the door.

Cleaning the engine   First remove the debris that you will find in the channels of the body, bonnet and the grill openings. The best way to accomplish this job is with an air line. Next, cover all electrical connections such as sensors, distributor and spark plug openings with either plastic film or bags. Use tape to seal the plastic surrounding these connections to prevent water from reaching them.

Completely wet down the wings, grill, top and bottom of the bonnet and the entire engine compartment. A pressure washer is ideal, but not essential. It is important to wet down all painted surfaces that surround the engine compartment, because the solution that you will use for cleaning may strip the protective coatings on these surfaces.

If the degreaser product is undiluted, carefully read the instructions regarding the ratio of product to water. A pressure tank sprayer or a spray bottle for application of the product is best. The engine cleaners work most effectively when all of the surfaces to be cleaned have received a thorough soaking. The product should be allowed to stay on the engine components for a few minutes but check the instructions on the container. Use of a brush on the engine and other surfaces may help to remove heavy oil deposits if you are not using a pressure washer. A steam cleaner is, of course, the ideal tool for this job if you have one available.

To remove the engine cleaner, completely soak down the entire compartment and the surrounding surfaces. If you use a pressure washer, be careful not to get the nozzle too close to any electrical connectors. When the compartment has dried, spray on a rubber and plastic conditioner.

Rubber and plastic   There are two causes of damage to exterior rubber and plastic surfaces:

- UV light
- ozone.

Most products that are designed condition plastic and rubber have silicone as an ingredient. However, there is some controversy regarding silicone as a component in both non-porous and porous surface conditioning. There are some people that believe that it will dissolve certain components in the rubber and cause cracking. It looks good when you first apply it though!

New car dewaxing   When a new vehicle is delivered from the factory, it is often protected with a type of wax or lacquer. This must be removed prior to the customer taking delivery. There are two main types of protection, often referred to as:

- hard wax
- soft wax.

An appropriate solvent is used to soften the wax. The vehicle is then washed in the normal manner, most often with a pressure washer or steam cleaner.

Exterior cleaning equipment

Pressure washers can increase the water pressure from about 4 bar up to 100 bar. This can actually tend to save water, as the washing process is much faster. Steam cleaners are similar except that they use high pressure and very high temperature water. The water is pressurised by an electric pump and heated by burning paraffin or a similar fuel. Steam cleaners are ideal for engine and chassis washing operations. Most allow some form of detergent to be added. Another popular use is for washing down new cars after a solvent has been used to soften the protective wax. Figure 14.4 shows a typical steam cleaner.

Figure 14.4 A good quality steam cleaner

Polishing tools are useful for exterior cleaning. They can save a lot of time when polishing or buffing up. An important precaution, however, is to take extra care that you do not damage the paintwork. Figure 14.6 shows some of the equipment needed for professional valeting.

LEARNING TASKS

➡ Look back at the key words. Explain each one to a friend, and/or write out a short description to keep as evidence.

➡ Examine a real car in detail and make a note of the exterior valeting procedures required.

➡ Write a short list giving advantages and disadvantages of using tools and equipment for exterior valeting.

Figure 14.5 Quality valeting products

3   Interior cleaning

Vinyl and leather

Before conditioning any surface, it must be completely clean. The first step in cleaning the interior of a vehicle is to give it a thorough vacuuming. The best type of vacuum will have a flexible extension with a reducing section or a small brush that can be attached. This type of device will allow you to reach into all of the areas that are difficult to reach in the dash and console areas.

After vacuuming, use a cloth and a spray bottle for the application of the cleaner. Be careful to test a small area of the surface before you begin, to make sure that the product will not produce any damage or significant colour removal.

The cleaner may not remove scuffmarks from the door panels and the plastic covers surrounding the base of the seats. However, there are products that will usually overcome this problem. These products are Acetone and Methyl Ethyl Ketone (MEK). You should be extremely careful with these products and never use more than a tiny amount followed by quick removal.

They should not be used in an enclosed or unventilated area.

Following cleaning, the conditioning step will be essentially the same process. The specific instructions for the product you have selected should be followed. In most cases the manufacturers will recommend buffing the conditioning product after it has had a few minutes to absorb into the surface. Regular cleaning and conditioning of vinyl and leather will extend the life of the product by a significant degree.

Carpets and upholstery

These cleaning products usually come in three versions:

■ spray bottles

■ aerosols

■ concentrates.

The most flexibility is obtained through the concentrates. The manufacturer will usually give instructions for how to dilute the product for several types of application. Again, care should be taken to assure that the product will not cause any unwanted result such as colour removal.

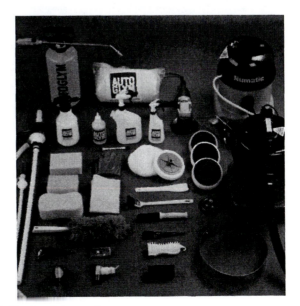

Figure 14.6 Tools of the trade

The surface should be thoroughly vacuumed both before *and* after use of a cleaner. The benefit of thorough re-vacuuming is that it will remove the contaminants that the cleaner has lifted from the surface. Most professional valeters will use a wet vacuum with a detergent in the water. The fluid is sprayed in to the upholstery under pressure and then vacuumed out.

Air fresheners There is a wide range of scents available in deodorants/air fresheners. An aerosol spray directly on the surfaces that will absorb the product is a good idea. These surfaces include the carpet, floor mats and the cloth upholstery (not vinyl and leather). Don't select products that have an overpowering aroma!

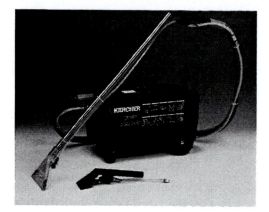

Figure 14.7 A typical 'wet-vac'

Interior cleaning equipment The main piece of equipment for interior cleaning is the wet vacuum cleaner (Figure 14.7). Water and suitable cleaning fluid is forced into the upholstery under pressure. A strong vacuum is then used to suck out the water and dirt. The water is collected in the machine for later disposal. Carpet cleaners used in the home are very similar. A normal, but heavy-duty, vacuum cleaner is also an essential interior cleaning tool.

Figure 14.8 Learn how to do the job properly

LEARNING TASKS

➡ Look back at the key words. Explain each one to a friend, and/or write out a short description to keep as evidence.

➡ Examine a real car in detail and make a note of the interior valeting procedures required.

➡ Write a short list giving advantages and disadvantages of using tools and equipment for interior valeting.

4 Operational checks and records

Introduction As with any other service or repair procedure on a light vehicle, records and checks are very important. Valeters can often be blamed for defects that were on the car before they started work. This is why they should be recorded and agreed with the customer. It is also important to record what has been done and when, as well as what the customer wants to be done and by when!

Defect recording procedures Most main dealers have their own procedures for recording defects and/or passing instructions to the valeter. Figure 14.9 shows a typical example. The pictures of the vehicle are used to mark on any defects. The customer is then asked to sign as confirmation that s/he agrees. Instruction can also be added to this sheet. Some very high quality and very expensive polish coatings are now available for vehicle use. They claim to seal the paintwork for a significant time, for several months if not years!

Valeting Instruction and Defect Report Sheet

Customer:

Name	
Address	
Phone	

Vehicle Details:

Make	
Model	
Reg. No.	

Instructions:

Wash only	PDI	Full Valet	Exterior	Interior	Engine	Other

Notes:

Defect Report Exterior: Defect Report Interior:

Customer signature: Date:

Valet carried out by: Date:

Figure 14.9 Valeting instruction and defect report sheet

Organisational procedures In common with all the work that companies carry out, each different organisation will have its own particular procedures. I can't even begin to cover all variations here, but I can stress that part of working as a team is to follow the set down procedures.

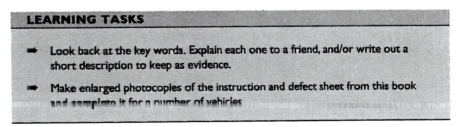

LEARNING TASKS

➡ Look back at the key words. Explain each one to a friend, and/or write out a short description to keep as evidence.

➡ Make enlarged photocopies of the instruction and defect sheet from this book and complete it for a number of vehicles.

15 Coaching individuals and key skills

1 Introduction

Start here! Key skills are the *'key skills'* you will need, on top of your technical knowledge and skill, to carry out your job to a good standard. One of the most important aspects is that these skills are transferable. In other words, they are useful skills to have for life in general as well as your work in the motor trade. I know it is hard at times to think ahead, but try to consider the number of further opportunities that will be available to you as your career and life progresses. Think of the benefits of being able to:

- communicate well
- use information technology
- apply and use numbers
- work with others
- improve your own learning and performance
- **and** solve problems.

As I said in the introduction to this book – put in the work now and it *will* pay off in the future!

Figure 15.1 Key skills are transferable to other jobs

Coaching individuals means that as you progress through your career, you will be expected to pass on your skills to others, in much the same way as others have passed on their skills and knowledge to you. The key issues or stages you will go through when coaching can be summarised as follows:

1. identify learning needs
2. consider the best way for the learner to learn
3. choose a suitable time and place, and decide what objectives need to be covered

KEY WORDS

■ All words in the table

4. use suitable language, cover the subject in small steps, and at the right pace for the individual

5. make the learner feel at ease

6. give constructive feedback

7. overcome any factors which may inhibit learning

8. check the learner's understanding and progress

9. allow application of the learning and match it against objectives

10. use different types of learning activity

11. encourage learners to recognise their own learning achievement.

Terminology

Key skills	Communication; Information technology; Application of number; Working with others; Improving own learning and performance; Problem solving
Life skills	In a way, these are very similar to the key skills above. I would suggest that everything that you experience and learn from improves your life skills
Self development	Always take the opportunity to develop yourself. When you have achieved something, the feeling you get is really good – not to mention the opportunities for better work and more money your development may achieve
Aims	If you play football, the overall aim is for your team to score a goal. The aim or goal of this book is to give you the UPK for your NVQ
Objectives	To achieve an aim you cover certain objectives on the way, but let's get back to football for an example. To score a goal your objectives (simplified a bit) may be as follows: Keeper throws to the left back, who crosses the ball to a midfield player, who passes to the right winger, who from the corner crosses to the centre forward, who heads the ball into the net. Each objective is like a step towards the aim or goal
Coaching	Simply the act of helping an individual or individuals to learn new skills
Learning methods	The way in which the learning takes place. You can for example, learn by reading a book, watching a demonstration or practising a task
Constructive feedback	Here is my constructive feedback to you: 'Well done for reaching this far in the book, keep up the good work! But don't forget to study any parts again that you are not sure about'
Planning	Think about a task before you do it, whether it is to do a job, coach others or whatever. Remember the six P's: Prior Planning Prevents P*** Poor Performance. I can't remember the fourth 'P' so I just put in stars. This may be due to lack of planning on my part!
Assessing	A way of making sure the learning is turning out how you intended. Maybe you have demonstrated how to change a drive belt. If your learner can then repeat the task you would assess him or her as being competent at this task
Review	Checking the progress of something or someone
Motivation	An incentive or reason for doing something. Maybe the reason we like to learn is because it's interesting, or because we may improve our pay and conditions as our work progresses
Performance criteria	Performance criteria state exactly what you must be able to do
Range	Subjects or systems you will cover during the performance
UPK	Underpinning Knowledge

LEARNING TASKS

➡ Look back at the key words. Explain each one to a friend, and/or write out a short description to keep as evidence.

➡ Write a short explanation about why transferable skills are useful.

2 Coaching individuals

Introduction

Coaching individuals simply means helping others to learn new skills and knowledge. It is, of course, not possible for me to cover every situation you will come across when coaching others. Likewise, I cannot cover every bit of automotive technology you will ever need. However, I can give you all the basic information and help you develop your skills in finding out what you need to know.

KEY WORDS

- Planning
- Feedback
- Step by step
- Opportunities
- EDP

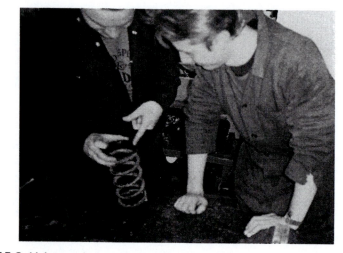

Figure 15.2 Helping others to learn is very rewarding

The knowledge and skills you will need to coach others will develop over time. You have probably learnt a lot already by watching others. Your reaction to the efforts of others coaching you may have been one of the following:

- 'That was a good way of doing it.'
- 'I wouldn't have done it like that.'
- 'What's going on?'

Either way you will have learnt something about the subject and about how you would coach others. In the next section, I will outline an example situation where coaching will take place. I will then, in the following section, apply the UPK requirements to this example.

Example situation

You have completed your NVQ to level 3 and are now in your fourth year of work for the same company. You are expected to carry out all normal jobs without supervision. You have your own work bay in the workshop and are expected to supervise and help to coach the new apprentice. Continuing at college one evening per week, you are developing your diagnostic skills further.

The learner started at the company about nine months ago and has been attending college for six months. The learner therefore knows all that s/he will ever need to know! The learner has started to carry out many tasks to assist with servicing work but wants to be able to do much more. S/he can tend to rush at things rather than being willing to ask you questions.

The company is a non-franchised garage carrying out work on all makes of light vehicle. It is a well respected company in the area, and charges customers a reasonable rate for a good job. Most specialist equipment is available for most tasks. Four other technicians work for the company, one is older and very experienced and the other three are at broadly the same stage as you. The boss has placed a lot of trust in you by allowing you to supervise the new apprentice.

The aim overall is that the learner should be able to carry out work to the company standards and complete his/her qualification. The shorter-term objectives are that the learner should be able to carry out more servicing-related work.

The task or objective is to coach the learner on how to jack up a car and change brake pads.

UPK required for coaching individuals

In this section I will use the example highlighted to help explain the UPK statements that follow. I have also included a step by step guide to this particular task at the end of this section. You should be able to adapt this to many other situations.

UPK statement	Interpretation of the example situation
How to identify learning needs	The needs in this case are that as well as having to learn more for his/her qualification, the learner must to be able to carry out further servicing related work to meet the needs of the company
Different styles of learning	We all learn in different ways but in this example you must choose the learning method most appropriate to the learner, the facilities, the task and the time available. Methods or styles available to you will vary, but it is likely that in most cases 'Explanation, Demonstration and Participation' (EDP) will be most appropriate. In other words: Tell 'em, show 'em and then let 'em have a go!
How to match coaching opportunities with individual learning needs and learning objectives	I have suggested just one task for this example, but it is likely that you will know of several tasks or have several objectives in mind appropriate to the needs of the learner. In this way, you can make use of naturally occurring opportunities such as changing brake pads on the vehicle you are servicing at that time
How to sequence and pace information and gauge appropriateness of language for individual learners	Try this: 'To *facilitate the extraction of the friction material, sequential operation commencing with hydraulically assisted support should be the primary accomplishment'.* Or this: *'To make it easy to get the pads out, work step by step starting with jacking up the car'.* The pace of your coaching should be appropriate to the learner's ability. It is better to take your time and work in steps that the learner will understand
How to put learners at ease	Try this: *'If you don't learn how to change pads quickly you will probably get the sack'.* Or this: *'It is important to learn this task but you must take your time and ask questions if you are not sure. You can practice several times if necessary but I'm sure you won't need to'*
How to give constructive feedback	We all like to be patted on the back from time to time. When a job is completed well, don't forget to tell the learner how well s/he has done. Equally important, however, is to be able to comment constructively on a job that was not completed satisfactorily: *'You did well on all parts of that job, except you let the calliper hang on the flexible brake pipe which could have caused damage. Remember this next time and you will be able to complete all of this task to a very good standard'*
Likely factors which inhibit learning and possible ways of overcoming them	This statement is really about communication. If the learner is finding it hard to understand a specific task, you will need to find a way of overcoming the problem. For example, the learner may have trouble interpreting instructions from the workshop manual. You could verbally explain what to do in more detail or spend more time practising this skill

UPK statement	Interpretation of the example situation
How to check learner's understanding and progress	Verbal questions are useful to check what you cannot infer from the learner's performance. For example, observing the learner changing pads is the main assessment but asking why the brakes should be cleaned with solvent and not an air line will help you get a full picture of their knowledge
Opportunities available for learner's to apply their learning and how to match them against learning objectives	As suggested earlier, it is likely that you will know of several tasks or have several objectives in mind for the learner. In this way, you can make use of naturally occurring opportunities. This could be changing brake pads on the vehicle you are servicing at that time, or ensuring that if the learners objectives are to be learning about brakes then you try to arrange for the learner to practice those particular jobs
How to integrate different learning activities	'Integrate' means to include the learning as a part of normal activities. The learner could be encouraged to work with the other technicians if they happened to be covering appropriate activities. For example, in this situation you could arrange for the learner to take part in jobs involving several different types of disc brakes
How to encourage learners to recognise their own learning achievement	It is always good to look back in time a little and compare what you know now with what you knew then. You should encourage the learner to do this, as it will help them realise how well they are progressing. In fact, why don't you think back a year or so and consider just how much more you know now than you did then? You can now begin to recognise your own achievements because I have just encouraged you to do so!

My step by step guide to coaching someone how to change brake pads (remember EDP):

1. Explain what the job is you are about to do.
2. Explain the *basic* principle of how brake pads work.
3. Explain the important safety issues: brakes are a major safety item on the car, jacking and supporting the car is important for your own safety and the possible risk to health from the brake dust.
4. Demonstrate how to jack up the vehicle and fit axle stands. Point out suitable jacking points if necessary.
5. Demonstrate how to remove the wheel on just one side of the vehicle.
6. Demonstrate how to remove the particular type of calliper on this car.
7. Explain briefly about other types of calliper fixings.
8. Demonstrate how to remove the pads.
9. Demonstrate how to clean the disc and calliper assembly.
10. Demonstrate how to push back the calliper piston.
11. Explain why you checked the fluid reservoir was not overflowing.
12. Explain why you checked that the pads are correct for the particular car.
13. Demonstrate how to grease the backs and edges of the pads.
14. Explain why special grease is used.
15. Demonstrate how to refit the pads and secure the calliper
16. Explain why you pump the brake pedal.
17. Demonstrate how to refit the wheel and torque up the wheel nuts.
18. Allow the learner to participate now by repeating all of the above on the other side of the car.
19. Observe the learner at work, giving constructive feedback.

20. Summarise the now completed task, allowing the learner to comment and ask questions.

I know this seems like a long list but if you look carefully you will see that each step on its own is quite easy to follow. This is a very important skill to learn when coaching others. Chop up the task into small manageable chunks; it is then both easier to teach and easier to learn.

LEARNING TASKS

➡ Look back at the key words. Explain each one to a friend, and/or write out a short description to keep as evidence.

➡ Consider several other tasks similar to the 'Brake pads' task described above. Write out step by step guides to each.

➡ Look at the eleven UPK statements again and write a few lines about each to explain them in your own words. Give examples to help your explanation.

Figure 15.3 You may be coaching others on many tasks

3 Key skills

Introduction Key skills are the basic requirements for almost any job. In fact, they are common to most jobs and are useful for life in general. They are not required for NVQ level 3, but must be completed if you are on a modern apprenticeship scheme. The qualifications are gained just like your NVQ, which is by building a portfolio of evidence. In many cases, your NVQ portfolio, with a little extra thought and work, could also contain most if not all of the evidence required for the key skills.

Key skills for modern apprentices The following is a list of the key skills you must complete if you are following the modern apprenticeship scheme. For each, I have given one or more simple examples to help you understand what is required. Your Assessor or Teacher will give you more details about exactly what you must do to gain the qualification. However, one secret is to integrate the key skills, or in other words, make them part of your main subject of study wherever possible. Remember my examples are just one small part and further work will be required.

Key skill Title	Level	Simplified example
Communication	2	You should be able to produce written reports, complete forms and take part in discussions constructively
Information technology	1	Use a PC at work to be able to look up customer details or search for parts
Application of number	2	Complete an invoice including VAT calculations. Be able to work out antifreeze percentages
Working with others	3	Form a vital and active part of a team; get on well with your mates, as well as with your boss and teachers!
Improving own learning and performance	3	Take responsibility for your own situation, work to learn more and strive to do your work better
Problem solving	3	Remember my stages of fault finding? 1. verify the fault 2. collect further information 3. evaluate the evidence 4. carry out further tests in a logical sequence 5. rectify the problem 6. check all systems. These can be applied to many situations with a little adaptation. Be able to think and act on your own but also know when to stop and ask for help.

KEY WORDS

- Communication
- Information technology
- Application of number
- Working with others
- Improving own learning and performance
- Problem solving
- Stages of fault finding

LEARNING TASKS

➡ Look back at the key words. Explain each one to a friend, and/or write out a short description to keep as evidence.

➡ Examine the evidence in your NVQ portfolio and consider just how much of it could be used towards the key skill qualifications.

16 Conclusion and what next?

1 Revision

Stop here!

KEY WORDS

- Revision
- Institute
- Internet
- Further study
- Money
- Prospects

Figure 16.1 Stop here!

If you have made it this far from the front of the book – congratulations! If you have not yet completed all your assessments, then I suggest now is a good time to work on a bit of revision. Remember the best way to revise is not to start again, but to look back at the *'Key words'* and *'Learning tasks'*. You will find, by now, that most of them seem quite easy. This is because you know the subject so much better than when you started. If some of the *'Key words'* and *'Learning tasks'* are still not clear, then a little further reading is needed in that particular area. Also, ask your Assessor or Teacher for some advice.

Now you are at the stage of completing a level 3 qualification, there is something else you should know! By the way, you should also be proud of your achievements. What you need to know now is that there is still a lot more to learn. Please don't think that this is the end of your studies. Technology is changing all the time – work to keep up with it! Lots of other materials and opportunities are available to you. Here are just some ideas:

- manufacturer's course
- correspondence courses
- evening class at college – C&G modules/BTEC higher qualifications
- video training
- other books – for example: 'Automobile Electrical and Electronic Systems' by Tom Denton!
- computer-based training
- press and magazines
- the Internet – see below for more details.

At this stage, you could also be thinking about joining a professional institute as a student member. For example:

- Institute of Road Transport Engineers – IRTE
- Institute of the Motor Industry – IMI.

Well done once again, I know by now you will understand a lot about motor vehicle technology and the industry in general. If you still enjoy learning about the subject, then that is even better. Keep up the good work!

LEARNING TASKS

- ➡ Look back through the book at the 'Key words' and 'Learning tasks'; study again the parts you are not quite sure about.
- ➡ Look back at the last two key words and think of the money and prospects!

Figure 16.2 Start again here!

2 The World Wide Web

Start again here! If you have access to a computer and modem then you are no doubt already interested in the Internet and the web. In this very short section, I just want to highlight some of the resources available to you on the Net. All the major vehicle manufacturers have web pages, as do most of the parts suppliers. There are user groups, forums for discussions and many other sources of information, or let's admit it – just good fun. So go ahead and surf the Net. I would be most pleased to meet you at my website where you will find up to date information about my books, new technologies and links to just about everywhere. Just point your browser to:

http://ourworld.compuserve.com/homepages/tom_denton/

To help further here are just a few interesting links:

http://www.topgear.com/	Top Gear
http://www.ferrari.it/	Ferrari
http://www.ford.com/	Ford
http://www.lotuscars.com/	Lotus
http://www.jaguarcars.com/	Jaguar
http://www.golucas.com/	LucasVarity
http://www.ev.hawaii.edu/	Electric Hybrid Vehicles
http://www.sae.org/	Society of Automotive Engineers

Good luck with your future studies and work, it's been good to know you . . .

Figure 16.3 Keep up the good work

Index

Ackerman steering   163, 164
Active safety   220
Adaptive cruise control   231
Advising customers   257
Air bags   222, 223
Air conditioning   232, 234
Air flow meter   80
Alarms   241
Alkaline batteries   191
Alternators   193, 196
Antilock brakes   181, 225
Anti-roll bar   149
Approval markings   204
Augmentation   256
Automatic transmission   131, 135, 137, 251

Batteries   188
Battery ratings   190
Beam patterns   202
Brake shoes   176, 179
Brakes   176, 253, 266
Bulbs   203

Camber angle   159
Carburation   74
Castor angle   160
Catalytic converter   117
Central door locking (CDL)   239
Charging   190, 193
Circuits   205, 206
Closed loop control   65, 184
Clutch   123, 250
Coaching   280
Constantly Variable Transmission (CVT)   135
Cooling   247
Crankshaft   38, 46
Cruise control   229
Cylinder liners   36, 45

Dampers   152
Diagnostics   243
Dial gauge   31
Diesel fuel additives   55
Diesel fuel injection   55, 98
Diesel fuel injection pump   56, 102
Diesel injection, electronic   103
Diesel injection, governor   103, 104, 105
Diesel injection, metering   103
Diesel injection, timing   103
Diesel injectors   57, 99, 100
Diesel smoke   58
Differential   127, 129
Diode   195
Direct ignition   71, 72
Disc brakes   176
Distributor   67

Distributorless ignition   71
Drive shafts   252
Drum brakes   176
Dwell map   91

Electric seats   238
Electric vehicles   218
Electric windows   240
Electrical systems   184, 254
Electronic carburation   76
Electronic control of transmission   136
Electronic diesel injection   103
Electronic ignition   67
Engine analysers   25
Engine balancing   43
Engine breathing   49, 52
Engine management   64, 248
Engine, 'K' series   44
Engines   35, 235
Evidence   6
Exhaust faults   247
Exhaust Gas Recirculation (EGR)   117
Exhaust gases   27, 118

Final drive   127, 252
Four-wheel drive   129
Four-wheel steering   169
Fuel cells   192
Fuel injection, petrol   78, 93
Fuel system faults   247

Gas discharge lamps (GDL)   207
Gauges   212
Glow plugs   58

Headlights   201
Health and safety   12
Heating   232, 233
Hot-wire fuel injection   84
Hybrid vehicles   219
Hydraulic power brakes   182
Hydro-pneumatic suspension   151

Ignition timing   66
In-car entertainment (ICE)   256, 258
Indicators   210
Inertia starter   198
Inlet manifold, variable   112
Instrument displays   215
Instruments   211, 215, 216
Interference suppression   261
Internet   288

Key skills   280, 285
Knock control   62

Lambda control   119
Lambda map   80

Lambda sensor   81, 113
Liquid Crystal Display (LCD)   216
Lead-acid battery   189
Light Emitting Diode (LED)   208, 215
Lighting   200
Load compensation, brakes   180
Lubrication   54

Master cylinder   178
McPherson strut   145
Micrometer   30
Mobile communications   260
Modern Apprenticeship   2, 7
Modifications   265
Mono-jetronic   87
Motronic   89
Multimeter   24
Multiplexed wiring   186

National Vocational Qualification (NVQ)   2, 3

Open loop control   65, 184
Organisational procedures   279
Overdrive   129

Passive safety   220
Personal Protective Equipment (PPE)   270
Petrol fuel injection   78, 93
Petrol fuel injectors   65, 82, 86, 114
Pistons   37, 39, 46, 47
Power steering, electric   167
Power steering, hydraulic   165, 167, 170
Pre-engaged starter   199
Pressure gauge   24
Programmed ignition   69
Pulse generators   68

Radar   232
Random Access Memory (RAM)   64
Read Only Memory (ROM)   64
Record keeping   279
Rectifier   195
Regulator   197
Roof rack   263

Safety   270
Safety policy   14
Safety systems   220
Screen heating   237
Security   241, 262
Self diagnostics   111
Sensors   212
Serial port   28
Slip angle   173, 174

Sodium sulphur batteries 192
Spark plugs 72
Spark plugs heat range 72
Split line brakes 177, 178
Springs 149
Starting 198
Steam cleaner 276
Steering 156, 165, 253
Steering characteristics 158
Steering geometry 157
Steering rack 165
Suspension 142, 252, 266
Suspension, front 144
Suspension, rear 146
Swirl 47
Swivel axis inclination 161
Synchromesh 122, 126
Systems 184

Tappets 48, 61
Test equipment 23, 244
Timing belt 52
Timing map 71, 91
Tools 20
Torque converter 122, 131, 133, 138
Torque curves 35, 39
Tow bar 267
Tow bar wiring 268
Tracking 162
Traction control 228
Transmission 121, 251
Transmission, semi-automatic 136
Trip computer 214
Turbo chargers 41, 43, 60
Tyres 171, 174, 253

Ultra-violet headlights 207

Unsprung mass 144
Underpinning knowledge (UPK) 9, 283

Vacuum servo 182
Valeting 269
Valve timing 40, 62, 115
Vehicle condition monitoring (VCM) 213
Vacuum fluorescent display (VFD) 217
Voltage stabiliser 212

Waveforms 32
Wheels 171, 253, 264
Wipers 208
Wiring 185
World Wide Web (WWW) 288